Plant Pathology

Plant Pathology

PRINCIPLES AND PRACTICE

D. Gareth Jones

Open University Press

Milton Keynes

Open University Press
Open University Educational Enterprises Limited
12 Cofferidge Close
Stony Stratford
Milton Keynes MK11 1BY, England

and

242 Cherry Street
Philadelphia, PA 19106, USA

First published 1987

British Library Cataloguing in Publication Data

Jones, D. Gareth
　　Plant pathology: principles and practice.
　　1. Plant diseases
　　I. Title
　　581.2　　　SB 731

ISBN 0–335–15157–4

ISBN 0–335–15154–X Pbk

Grateful acknowledgement is given to BASF (UK) Ltd for the colour photograph of
Botrytis cinerea *on grapes, shown on the front cover.*

Text design by Clarke Williams
Printed in Great Britain at the Alden Press Ltd, Oxford

To Anita, Huw, Philip and Justin

Contents

PART II

List of figures and tables

List of figures

List of Plates

Compendium

List of Tables

Preface

There is certainly no shortage of books on various aspects of the subject of plant pathology, as the list of recommended books in the bibliography testifies. However, after over twenty-five years teaching plant pathology, students repeat the same question year after year, 'Is there one book which would give me a reasonable coverage of the subject as a whole?' The answer has always been in the negative and only the readers of the present book can judge whether or not the situation has changed.

It may well be that my objective was unrealistic in a book of this size. Inevitably, there will be criticism of insufficient mycology, too much by way of individual disease details or not enough on the physiology of the diseased plant. What was required in an introductory text and how much of each topic should be included? My choice of chapters and the depth of their content was arrived at as a result of my teaching experience. Graduates of the Department of Agricultural Botany at University College of Wales, Aberystwyth, will immediately identify with this book as it reflects the teaching courses presented by myself and my long-standing friend and colleague, Professor Ellis Griffiths.

The book is aimed primarily at first and second year undergraduates of plant pathology, agricultural botany or applied biology. Colleges of higher education, farm institutes and especially the agricultural colleges should also find the book of value if their courses have any inclination towards crop protection. Without apology, the book will not match the detail of others written on specific topics. For the honours student who is required to, or the reader who wishes to, pursue particular aspects in greater depth, specialist books are available and referred to in the list of books for background reading at the end of the bibliography.

My experience in jointly writing a specialist book *Cereal Diseases—Their Pathology and Control* with Brian Clifford, greatly helped me with the format and style of the present book. Sufficient detail is given to counteract criticisms of superficiality and key references, especially of some excellent reviews, provide additional reading. It is hoped that the reader is provided with a clear insight into the principles of plant pathology, the biology of the host–parasite interaction and the methodology of study. The compendium of diseases has been carefully chosen to be of global interest and to provide examples of the various management strategies. With the considerable emphasis placed on epidemiology and current chemical control methods, it is hoped that the reader will truly learn of the subject of plant pathology—its principles and practices.

Acknowledgements

I am pleased to record my thanks to many friends and colleagues, commercial companies and government institutions. I am particularly indebted to Professor Ellis Griffiths for his critical and constructive comments during the preparation of this manuscript and to Professor Hugh Rees FRS and Dr Gareth Evans for their helpful encouragement and valuable advice.

The writing of this book would not have been possible without constant recourse to the published results of the research of many hundreds of plant pathologists, past and present. I acknowledge my debt to these publications here and explain that only a selection of references are included in the text and these may be used to supplement the detail provided.

Many photographs and figures were kindly provided by many sources. These are acknowledged individually where included, but I am most grateful for the collaboration of Drs A. Brunt, T. L. Carver, J. Cotten, C. M. Garrett, J. Gibbs, J. E. King, G. A. Mathews, R. A. Noon, R. W. Polley, D. G. A. Walkey, J. M. Webb, P. J. Whitney and Prof. G. F. Pegg. In addition, I am most grateful for the help of Prof. R. C. Pearson, Cornell University, USA and Prof. D. A. Roberts of the University of Florida, USA.

Many agrochemical companies readily contributed detailed information on the nature and use of the many fungicidal and bactericidal products now in use. I am particularly indebted to ICI Plant Protection Ltd, BASF (UK) Ltd, Bayer (UK) Ltd and Schering Agriculture Ltd. Other technical information was readily supplied by E. Allman and Co. Ltd, Chichester, Burkard Manufacturing Co. Ltd, Rickmansworth, and Micron Sprayers Ltd, Bromyard. I would also like to pay tribute to the cooperation of many ARFC institute directors and liaison officers and to the MAFF and a great number of its personnel.

Lastly, I am pleased to acknowledge the help and support of the staff of Open University Press, Mrs E. Kenny and Mrs J. Ebenezer for secretarial work and, last but not least, my patient wife Anita for her help in checking and editing the manuscript.

PART I

1

The Disease Problem

The Past

Man's requirement for a plentiful food supply has always been threatened by droughts, floods, pests, and diseases. The problem of plant disease has been recorded in the very early writings, the Greek philosopher Theophrastus being credited with certain observational comments about cereal and legume diseases some three centuries before Christ. He wrote in ignorance of the causal organisms involved although he was astute enough to recognize host variation in susceptibility and the overriding effect of environment.

Over the 2000 years since Theophrastus, there was little or no progress in Man's understanding of plant diseases, although the intensification and increasing sophistication of his cropping was bringing more frequent disease calamities, and considerable pressure was being exerted to obtain more information about the pathogens involved in order to attempt to reduce their impact through some control measure.

Leeuwenhock's use of the compound microscope and his discovery of bacteria in 1675 introduced science into what had been hitherto a purely descriptive subject.

The death of some 20 000 troops of Peter the Great, Czar of Russia's army in 1772 was attributed to their eating rye flour which contained milled ergots, sclerotia of the fungus *Claviceps purpurea*. A better understanding of the disease might have altered the course of history, but the catastrophe occurred without immediate explanation unlike a similar tragedy in France in 1951 when contaminated flour was knowingly sold with a resulting four deaths and thirty-two cases of insanity.

The devastating epidemics caused by late blight of potatoes (*Phytophthora infestans*) in Ireland in the 1840s was an even greater tragedy causing the deaths of over one million people and resulting in over two million people emigrating, mostly to the United States. Again, the victims and the authorities were unaware of the real causes and it was 1857 when Speerschneider finally identified the fungus responsible for the potato famine.

The name of De Bary is inextricably linked with the proof of pathogenicity in many plant diseases, working initially with the smuts and rusts, then the downy mildews which included late blight of potato and finally with *Sclerotinia* diseases. De Bary was to plant pathology what Pasteur and Koch were to animal diseases.

In the late 1870s, the vineyards of Europe were seriously threatened by the downy mildew pathogen (*Plasmopora viticola*). The European vines had not been subjected to this disease previously but it had been inadvertently introduced from the United States. The rapidly spreading disease focused the attention of many agriculturalists of the day but none more than Professor Millardet of the University of Bordeaux who, in 1885, discovered the fungicidal properties of copper sulphate and lime by his observation that vines which had been sprayed with this sticky blue-white mixture were relatively free of downy mildew infection. This chance observation became a landmark in plant pathology for not only did it introduce to the world what must rank as the world's most famous fungicide, namely Bordeaux Mixture, but also it heralded the start of the fungicide industry as it is known today as commercial firms began screening chemicals for their fungicidal and bactericidal activity.

Another landmark in the social history of plant pathology is most certainly the enforced change in the drinking habits of a nation. Prior to 1870, Sri Lanka (or Ceylon as it was then) was an important coffee-exporting country, mainly to satisfy the British nation which was, at that time, primarily a coffee-drinking nation. Unfortunately, by 1886 the coffee rust pathogen (*Hemileia vastatrix*) had become so damaging that bankruptcy and disillusionment had reduced the coffee industry to negligible proportions. In its place, the planters turned to the tea crop and, within a very few years, Britain became a nation of tea drinkers.

The Present

During the present century, interest in plant pathology has increased dramatically, almost every country teaching the subject as a science at its colleges and universities and every government employing plant pathologists, often in considerable numbers, for research, extension and legislative tasks. This increased input into the science has resulted in important developments in many aspects of the subject, especially in respect of our knowledge of the physiology of disease, the use of epidemiological information for disease forecasting, the genetics of virulence in the pathogen and resistance in the host plant, and in the precision with which control treatments are applied.

The need for increased food production, the introduction of mechanization, the better understanding of the causal organisms of disease and the initiation of extensive crop improvement programmes incorporating disease resistance selection, have all contributed to the current awareness of the importance of plant pathology and the enormous losses suffered globally as a result of disease. Any figures for crop losses are always questioned and often reflect the interests of the statistics compiler rather than the true situation. However, even allowing for exaggerated estimates, there can be no doubting the seriousness of many of our crop diseases on a world scale, and statistics, unfortunately, are frequently superfluous on a local scale when a farm crop is badly ravaged. The economics of disease is, perhaps, best left to the professional economists as the pathologist will be confused by and argue with their definitions of 'normal yield' which would be taken as the baseline for calculating yield losses. In addition, there will be other equally confusing factors for it is an accepted fact that, in many crops, the conditions that are conducive to disease are also favourable to high yield. Such conditions are often encountered with the potato crop where the so-called 'blight years' are also normally good yield years. It should also be remembered that financial losses to the farmer can occur as a result of measures taken to combat disease as well as from direct disease damage. The farmer may have to buy more costly seed, a resistant cultivar perhaps, invest in costly spraying machinery, storage space, refrigeration and possibly specialized transportation vehicles as well as having to bear the cost of purchasing and applying chemical control treatments.

Plate 1.1   The use of 'tramlines' in cereal growing. (Photo: BASF (UK) Ltd)

Changes in Agricultural Practices

There can be no doubting the effects that the changing agricultural practices over the past century have had on the incidence and severity of plant disease. Nowhere has this been better illustrated than with the cereal crop. Traditionally, the cereal crops were the source of energy for the muscle power of man and his animals. Only sufficient was grown locally for his own needs. As man changed from the nomad to the cultivator, with the coincidental domestication of his animals, the nature of cereal cultivation changed. It became the dominant agricultural crop over vast areas, the corn belt of the United States for example, often with thousands of hectares being devoted to the monoculture of a single cultivar of wheat, oats, barley or maize. The consequences of such practices have been well documented in most plant pathology textbooks and the epidemic of Southern corn leaf blight in the United States in the 1970s is ample testimony to Man's neglect of the information obtained after decades of plant pathological studies.

The twentieth century heralded a new era in crop protection as more and more chemicals were developed which could be used for disease control. The advent of the highly specific and effective systemic fungicides in the 1960s even changed the emphasis of cereal cultivation. The inclusion of 'tramlines' (Plate 1.1) in a wheat or barley field is an extremely common sight these days, the implication being that the farmer is seriously commited to the application of chemicals, whether they be for weeds, pest or disease control, a philosophy which would not have been contemplated twenty years ago. Chemicals have certainly made their presence felt in many agricultural systems, but the overall view of plant pathologists should still be toward better management practices that would include hygienic measures plus the use of resistant cultivars where possible. Such an integrated approach is thought likely to be the most efficient way of stabilizing yields with each individual management practice increasing the efficiency of the others and, at the same time, acting as insurance policies against the failure of a single treatment.

The aim of both plant breeder and pathologist today is to shift the balance of the host: pathogen combination in favour of the host. Primarily, this is being achieved through the incorporation of resistance genes by breeding. However, experience during the last thirty years or so has emphasized the extremely ephemeral nature of some types of resistance (see Chapter 6) with new pathogen virulences appearing to overcome the introduced resistance genes in all too short a time. Again, management and breeding strategies such as the use of cultivar diversification and multilines have been developed to counter the erosion of resistance in the 'boom and bust' cycle and only the future can tell which approach will be most successful.

2

The Causes of Plant Disease

Introduction

Disease is usually defined as being a significant and harmful deviation from the normal physiological process of the plant. For the purpose of this text, a plant will only be considered to be diseased when visible symptoms are apparent. Some disease symptoms may be less obvious than others (see Chapter 3) and the effect upon the plant may be one of a quantitative or a qualitative reduction in potential yield. Disease diagnosis is thus of paramount importance but, it should be emphasized, the symptom picture may not be simple. Many symptoms may be common to several and very different causal agents—for example, wilting can be the result of a vascular pathogen or simply the lack of water. In addition, it is not uncommon to find plants affected by more than one pathogen and, although their symptoms may be separately distinguishable, their effects upon the plant may be impossible to evaluate accurately. Disease symptoms are important in the identification of causal organisms but their expression may vary considerably depending upon the natural resistance of the host, its nutritional history, its stage of growth and, most importantly, the environment. The host, the pathogen and the environment are often described as the **disease triangle** with each component being able to interact with the others.

Diseases can be caused by either pathogens or non-pathogens. Fungi, bacteria, viruses and nematodes are by far the most numerous and important infectious agents of plant disease but parasitic higher plants such as dodder, mistletoe and witchweed also contribute to a lowering of plant growth efficiency. In the global definition of disease, mention should also be made of the damage caused by pests such as insects, mites, slugs, snails, wireworms and even moles and mice.

Non-pathogenic diseases include a number of physical and chemical factors. Such diseases are often grouped together as 'physiological disorders' and include mineral deficiencies and toxicities, chemical pollutants, extremes of environment such as frost or drought and such modern-day technological mishaps as 'spray-drift' of pesticides on to a 'non-target' crop.

No matter what the crop, imbalance of essential nutrients in the soil will result in some form of physiological disorder. Nutritional imbalance may be due to either deficiency or excess and, in the case of the soil–acidity complex, related to soil pH, available/exchangeable calcium, the soil condition and thus the physiological condition is indirectly related to the primary problem. As mentioned above, real problems exist where a combined nutrient imbalance/disorder/deficiency occurs particularly for diagnosing by visual symptoms.

Fungal Pathogens

Scientific proof of the pathogenicity of fungi is a relatively recent event. In 1807, Prévost was able to conclusively show that the bunt disease of wheat was caused by a fungus (*Tilletia caries*) and that the presence of the fungus was the cause and not the effect of the disease. Other similar discoveries quickly followed, the events being not only biological landmarks but also marking the start of the study of plant pathology as a scientific subject.

Fungi are mostly microscopic, classified in the plant kingdom but devoid of chlorophyll. Their vegetative form is normally a thread-like **hypha**, or **thallus**, which often aggregates as a colony within its host where it is collectively termed **mycelium**. Characteristics of growth form, morphology, or reproductive organs and propagules form the basis of taxonomic classification (see Chapter 3) and a study of any order, family, genus or species will reveal infinite variation. However, it should be pointed out that, in general, the taxonomic positions of plant pathogenic fungi are not correlated with the kinds of diseases induced by those fungi. For example, *Phytophthora infestans* causes a foliar and tuber disease of potatoes known as late blight, whereas *P. cactorum* has been recorded on over forty families of flowering plants causing a variety of diseases such as damping-off, root-rot and fruit rot, and *P. palmivora* causes a pod-rot and canker of cacao.

Hyphae may be divided up, **septate**, into separate cells, or be elongated, undivided, **aseptate** tubes. Most fungi reproduce by producing spores and these can either be sexually or asexually produced. The most commonly produced spores are the asexual **conidia** which may be produced directly from the vegetative mycelium or from relatively simple subtending structures, the **conidiophores** (Plate 2.1). Sometimes, groups of conidiophores are grouped together to form a compact **coremium** or they may be aggregated into a more complex fruiting-body such as the saucer-shaped **acervulus** or the flask-shaped **pycnidium**. Conidia are produced as a result of simple mitotic cell division and, consequently, are genetically identical to the parent cells. Conidia are usually produced in large numbers and are the propagules of dissemination as well as providing adequate chances for the survival of the fungus; a review of fungal spore dispersal has been produced by Ingold (1971). Conidia normally account for the rapid build-up of disease in crops and may be produced continually over a long period. A good example of the mass production of asexual spores is the production of conidia by the cereal powdery mildew pathogen, *Erysiphe graminis*, which can be seen to rise from the crop in grey spore clouds if the foliage is disturbed by wind. Another form of asexual spore is the **sporangiospore** which is produced in numbers within a membranous bag or **sporangium** at the tip of hyphal branches. Many sporangiospores are motile with one or more flagella when they are called **zoospores**. Yet another asexual spore is the **chlamydospore** which is produced as a thick-walled spore within hyphal or conidial cells and are well adapted for long-term survival.

Sexual spores such as **oospores, ascospores** or **basidiospores** provide the fungus with increased chances of survival by the production of new genetic variation through meiotic recombination in the course of sexual reproduction. Sexual spores, or quite often the fruiting bodies in which they are produced, can often overwinter in the absence of a host plant, many having the ability to survive several years in a resting spore form. Resting spores of *Synchytrium endobioticum*, the causal organism of wart disease of potatoes, can survive up to 30 years in soil and many oospores of *Phytophthora* spp. are known to have

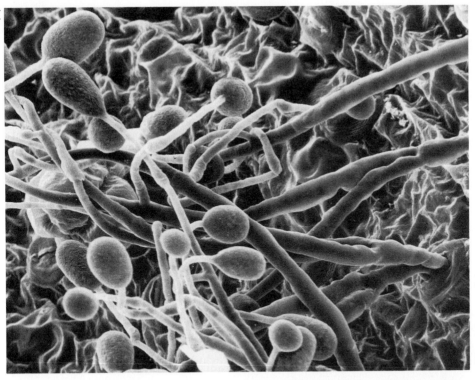

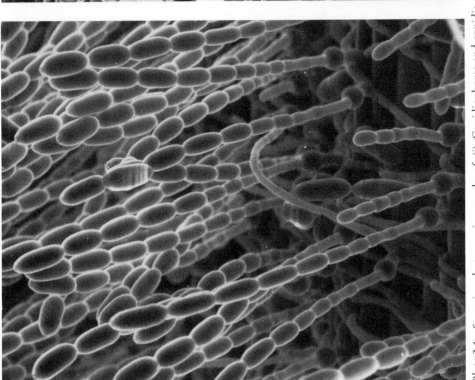

Plate 2.1 Scanning electron micrographs of: (i) Conidiophores and conidia of *Erysiphe graminis* (ii) Sporangiophores and sporangia of *Phytophthora infestans*
(Photos: (i) T. L. Carver, WPBS; (ii) ICI Plant Protection Ltd)

a longevity of several years.

As plant pathogens, fungi can be highly specialized not only in terms of the genus, species or even cultivar they can infect but also as regards the host tissue they colonize. They can be **biotrophic**, obtaining their necessary nutrient requirements from living cells in the hosts as, for example, the rust fungi; or they may be **necrotrophic**, pathogens which kill the host cells, often in advance of their colonization and live off the breakdown products of the dead cells. Fungal plant pathogens can also be conveniently divided into two groups depending upon whether they can live in the absence of the living plant. Those fungi that normally grow and reproduce only on living hosts have, traditionally, been classified as **obligate parasites**. This category of pathogens poses a problem to an investigator in that its members cannot be grown on artificial media such as the various nutrient agar media commonly used for the culture of micro-organisms. The remainder of fungal pathogens and all the bacteria should therefore be classified as non-obligate parasites, having the ability to live on both living and dead host tissues. Of these, fungi with a strong parasitic habit have been termed **facultative saprophytes** whilst the weakly parasitic fungi are called **facultative parasites**. The facultative pathogens are often not as specialized as the obligate pathogens and they have the advantage of being able to be cultured on artificial laboratory media.

The specificity of fungal pathogens in terms of host range has very important practical implications. In the 'highly specific' category are the rusts and powdery mildews which are, by far, the most important pathogens of cereals. Of world-wide significance is the black stem rust pathogen (*Puccinia graminis*) which only attacks members of the Gramineae. Subdivisions have been made within this species to distinguish between forms which are adapted to the various host genera. These *formae speciales* (single, *forma speciales*) are given the nomenclature of the host species predominantly attacked. The form attacking wheat is called *Puccinia graminis tritici* whilst *P. graminis avenae* attacks oats and *P. graminis secale* attacks rye. However, these subdivisions are not absolute and there is a limited amount of promiscuity over the host range.

The degree of specificity becomes even more precise with the further subdivision of each adapted form into **physiological races**, each race having a virulence specific for a defined spectrum of host cultivars. The relationship of each physiologic race to its potential hosts depends upon the compatibility of the genetic systems in both host and pathogen. The evidence suggests that the two systems have developed in parallel with a mutation to virulence in the pathogen being countered by the selection, either in nature or by the plant breeder, of hosts with the necessary genes controlling resistance against this virulence. Such host–pathogen relationships led Flor (1956) to put forward his **gene-for-gene** theory which is discussed elsewhere in this book (Chapter 6).

Another form of specificity concerns the host tissue attacked by the pathogen. Many root-infecting fungi never infect the foliage, a good example being the take-all pathogen (*Gaeumannomyces graminis*) whilst the ergot pathogen (*Claviceps purpurea*) is restricted to the floral organs.

During evolution, the fungi have become well adapted for widespread dispersal (Ingold, 1971). With various types of aerial conidiophores, spores which are often resistant to desiccation and radiation, the pathogen propagules are ideally suited for survival and long distance dispersal. The possession of resistant spores and often an active saprophytic stage further increase the perennation of many fungi and are important factors in the epidemiology of necrotrophic plant pathogens.

Bacterial Pathogens

In view of the importance of bacteria as pathogens of man and animals, it is perhaps surprising that they are not as important globally as fungi or viruses as agents of disease in plants. One possible reason for this is that bacteria tend to be thermophilic and, whilst they can be extremely damaging to plants in the tropics, they appear to be less well adapted to the overall climate of the temperate regions, especially the cold winter months. However, there are many important bacterial diseases of plants and these are described in very few textbooks, the most notable being by Dowson (1957), and as chapters in more comprehensive pathology texts such as Walker (1969) and Agrios (1969).

Bacteria are very primitive organisms with no nuclear membrane or mitotic apparatus, as such they are classified as **prokaryotes**. The bacterial cell wall is fairly rigid but sufficiently permeable to allow the inward passage of nutrients and the outward passage of waste matter, digestive enzymes and other excretory products. Most bacteria are unicellular and of varying shape. Some are spherical (cocci) others rod-shaped (bacilli), whilst others may be spiral (spirilla) or common-shaped (vibrio). However, all plant pathogenic bacteria are rod-shaped and some are capable of forming resistant spores which are common in animal bacterial pathogens.

Plant bacteria are motile by means of variously sited **flagellae**. These organs of motility may arise all over the cell surface, **peritrichous**, or may arise at one or both ends, **polar**. Many bacteria seem to grow best in a slightly alkaline medium whereas the cell sap of most plants tends to be somewhat acid (pH 5.6) Such characters as cell morphology, nature and number of flagellae are diagnostic features used in the classification of bacteria (see Chapter 3) but many other characteristics such as nutritional requirements, staining differences and serological reactions need to be included also.

Unlike most fungi, the identification of bacteria require that they are first isolated and purified in pure culture. On solid agar media, colony form can vary greatly and the points to note are colour, form and shape (raised, flat or in a film), surface (smooth, matt, glistening, rough), and margin (lobed, spreading or entire). Each bacterial colony can originate from a single cell (colony forming unit, c.f.u.) which multiplies very rapidly by simple division (**binary fission**) often with a generation time of from 20 minutes to 2–3 hours. The result of such repeated cell division is a colony of identical cells which, in the absence of mutation, breed true and maintain constant characters from generation to generation.

The bacterial nucleus consists of a single, much convoluted circle of DNA and has no nuclear membrane. The genetics of many bacteria have been widely investigated using methods involving **conjugation**, in which cell fusion is followed by recombination, **transformation**, where the genome of one bacterial species is altered by the addition of DNA from another species and **transduction**, where bacteriophage is used to transfer fragments of DNA from one bacterium to another. The recent discovery of plasmids in many bacterial genera has upset the traditional thinking of bacterial genetics. Plasmids are small, circular strands of DNA which are known to control certain fundamental processes in a range of bacteria. Most of the nodulation and nitrogen fixation genes of the symbiotic *Rhizobium* genus are known to reside on a single plasmid of around 150–400 megadaltons (MD). The gall-forming *Agrobacterium tumefaciens* owes its pathogenic character to a small region (*T*-DNA) on a single plasmid, the tumour-inducing plasmid (*Ti*-plasmid).

Almost all plant pathogenic bacteria have at least a brief saprophytic phase in the soil in addition to their parasitic phase on the host plant. At one extreme is the ubiquitous soil bacterium *Erwinia carotovora* which is responsible for causing a soft and very damaging rot on a wide range of hosts. This bacterial pathogen lives mainly in the soil and its translation into a plant parasite is both brief and erratic, depending upon the presence of wounds in the hosts and conducive soil and environmental conditions. In contrast, the fire-blight pathogen, *Erwinia amylovora*, is primarily a plant parasite whose numbers in the soil rapidly diminish in the absence of infected hosts. *Agrobacterium tumefaciens* is intermediate between these two extremes, the soil phase population declines very gradually after release from the plant but still presents a disease hazard for several years.

With very few having a resting spore stage, bacterial plant pathogens survive in or on some seeds, on plant debris or saprophytically in the soil. They possess no specialized disseminatory organs and rely for spread on other agencies such as water, insects, animals and man and, even after arrival on a potential host, they often require some wound to. facilitate entry as they do not possess the ability for direct penetration. Entry can also occur through natural openings such as stomata, hydathodes, lenticels and nectaries and, once established in the host, the bacteria live and multiply and cause damage leading to a variety of symptoms often leading to plant death. Bacterial diseases can be grouped into four categories based simply on the nature of the symptoms and the host tissue affected; (i) **parenchymatous, vascular, systemic** and **hyperplastic**. These groups are discussed elsewhere in this book (Chapter 3). There are only six genera of importance as plant pathogens, *Corynebacterium* and *Streptomyces* are the only gram-positive genera with species such as *C. sepedonicum* producing a ring-rot in potatoes. *Agrobacterium*, *Erwinia*, *Pseudomonas* and *Xanthomonas* are all gram-negative causing a wide variety of symptoms from soft-rotting to leaf spots and galls. *Streptomyces* is represented by only one species, *S. scabies*, which causes common scab of potatoes, but it should be said that this organism produces slender threats and not conventional individual cells and, whilst classified in the Actinomycetes, it has recently been acknowledged as a bacterium on the basis of its undoubted prokaryotic characteristics.

Viruses, Mycoplasmas and Rickettsias

Many definitions of the word *virus* have been suggested over the years, changes often being required because of the acquisition of new knowledge which helped to more clearly distinguish a virus from other disease agents such as viroids and mycoplasmas. A very recent definition by Mathews (1981) is very comprehensive and suitably explanatory:

> A virus is a set of one or more nucleic acid template molecules, normally encased in a protective coat, or coats of protein or lipoprotein, which is able to organize its own replication only within suitable host cells. Within such cells virus production is (a) dependent on the host's protein synthesizing machinery, (b) organized from pools of the required materials rather than by binary fission, and (c) located at sites which are not separated from the host cell contents by a lipoprotein, bilayer membrane.

Symptoms of virus diseases have been known for several hundreds of years. The most notable of these were flower-breaking in tulips which even achieved fame as subjects for paintings by the Dutch masters in the seventeenth century. However, it was not until 70 years ago that the cause of such conditions was identified as a virus.

The plant viruses consist of single strands of nucleic acid which are encapsulated within protective protein coats or shells (capsids). The nucleic acid is RNA with only cauliflower mosaic virus (Caulimovirus group) being a DNA virus with a double strand (ds) and maize streak virus (Geminivirus group) with a single (ss) DNA strand. The capsid is composed of many individual protein subunits (polypeptide chains) the arrangement of which varies with virus type. Virus particles are too small to be seen with the ordinary light microscope and are of various shapes (Plate 2.2), (a) **rod-shaped:** either rigid or flexible rods of up to around 300 nm long—includes potato viruses, S, X and Y and beet yellows virus; (b) **icosahedral:** a polyhedron with 20 faces—includes turnip yellows mosaic, cucum-

(i)

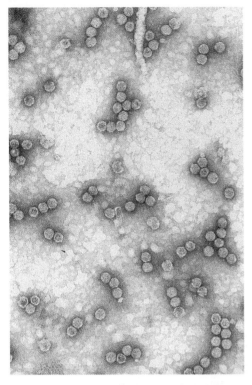

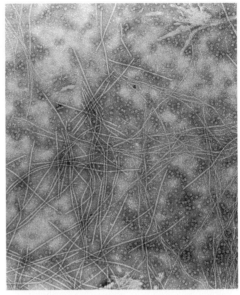(ii)

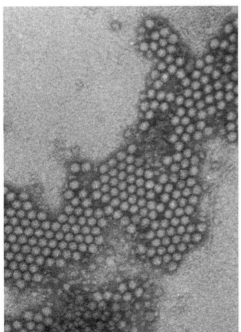

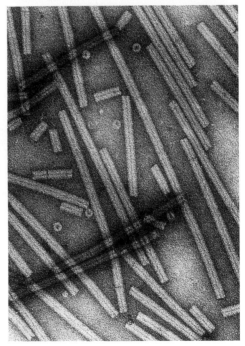(iv)

(iii)

Plate 2.2 A selection of virus particles photographed using an electron microscope
  (i) Tobacco ringspot virus
  (ii) Turnip mosaic virus and cucumber mosaic virus
  (iii) *Chenopodium* necrosis virus
  (iv) Tobacco mosaic virus
(Photo: National Vegetable Research Station)

Plate 2.3 Healthy mushroom (left) and the virus infected sporophores (centre and right). (Photo: R. G. Barton, GCRI)

ber mosaic and tobacco necrosis viruses; (c) **bacilliform:** smaller than most rod-shaped viruses (up to 60 nm long) and differs in that the ends of the particle are covered by protein subunits.

Viruses are also known to infect other classes within the plant kingdom and have become important as damage-causing agents of the cultivated mushroom, *Agaricus bisporus*. Mushrooms (Plate 2.3) can be so badly diseased that the whole crop is 'written-off' and destroyed. These fungal viruses, or **mycoviruses**, have mostly ds-RNA genomes and are frequently transmitted during spore production. Virus particles have been identified in a number of fungi including the take-all pathogen, *Gaeumannomyces graminis* and *Helminthosporium victoriae*.

Many diseases of plants were originally ascribed to infection by viruses even though no virus particles had ever been observed by electron microscopy. However, in 1967 the first report of the discovery of **mycoplasma-like organisms** (**MLOs**) was published in Japan in respect of the mulberry dwarf disease. Other reports quickly followed confirming MLO in clover phyllody disease, aster yellows and many other diseases previously described as virus diseases.

Mycoplasmas are pleomorphic in shape, have no cell wall other than a surrounding membrane. They contain ribosomal DNA and DNA in the form of a coil in their nuclear region. As infecting agents, mycoplasmas are usually confined to the phloem or xylem cells and are transmitted by leafhoppers and other sucking insects.

Other diseases, previously ascribed to virus infection or left as of uncertain origin, have now been identified as being caused by **rickettsia-like organisms** (**RLOs**) which resemble MLOs to some extent except that they do possess a cell wall. MLOs normally range in size from 0.1 to 1.0 μm in diameter whereas RLOs measure between 0.4 μm in diameter with a length of about 3 μm. RLOs have been associated with Pierce's disease of grapes, phony disease of peach and clover club leaf. Although very small, RLOs can be observed in the electron microscope, they are vector transmitted mainly by leafhoppers but the clover club leaf organism is transmitted by *Agalliopsis nonella* in which it can multiply and pass from generation to generation in the eggs.

Virus Symptoms

As can be seen from an examination of the common names of virus diseases, most are named after the type of symptoms they cause. Field symptoms are most important in disease diagnosis (see Chapter 3) and, of course, there is usually a good relationship between their severity and yield loss.

With the rapid multiplication of the virus particles in the infected host, changes are brought about in plant growth, form and function. Some changes may only be apparent with the aid of the electron microscope, but others produce internal and external effects that are readily visible and comprise the disease symptoms.

As with many plant diseases of fungal and bacterial origin, infection by viruses often causes a debilitation which manifests itself as a **stunting** or **dwarfing** symptom. The stunting may be of a general nature or it may be confined to specific plant parts or organs. It may be difficult to appreciate stunted growth in a uniformly infected crop but comparison with healthy individuals makes such plants very conspicuous. It should also be remembered that stunting can also affect the roots and these should also be examined in any appraisal of disease severity.

Colour changes are often associated with virus infection. These often result in **mosaic** patterns on leaves, stems or fruits. The mosaic symptom is due to a reduction in chlorophyll content resulting in yellow mottling, flecking, spotting, blotching or striping against the normal green background (Plate 2.4). Other colour changes are the 'break' symptoms in tulips already mentioned. These virus-induced symptoms have added to the commercial value of these particular stocks and have several forms, a streaking, flecking or sectioning to give a dual-coloured corolla.

Chlorosis and **necrosis** occurs frequently with many plant virus infections often starting as interveinal chlorosis and, in its most severe form, with extensive necrosis as in the internal leaves of cabbage infected with turnip mosaic virus or, together as chlorotic and necrotic concentric rings of tissue as in the ringspot symptom produced by the same virus on cabbage.

Virus infections also causes **leaf distortion,** as in the classical tobacco mosaic virus disease of tomatoes, or **stem distortion** as in the swollen shoot disease of cacao in the tropical regions. The latter disease is an example of abnormal cell proliferation in the host, probably due to changes in hormone concentration. Similar imbalances can cause **tumours** on leaves and roots, a good example being the 'warty' outgrowth, or **enation**, produced on pea leaves by the pea enation mosaic virus.

Virus Transmission and Infection

The fact that viruses are non-motile and have no adapted disseminatory characteristics implies that dissemination can only be achieved through the agency of vectors or by means of seed transmission and vegetative propagation of the host species. Laboratory procedures for the transmission of viruses tend to be atypical of natural methods, the most common technique being the application of plant sap from an infected to a healthy host possibly accompanied by minor wounding with a mild abrasive, such as carborundum. This **mechanical** or **sap transmission** is probably of infrequent occurrences in nature although potato virus X can be transmitted by leaves rubbing against one another (Roberts, 1946). In addition, Broadbent (1976) has shown that tomato seedlings can become infected through the roots which come into contact with debris from tobacco mosaic virus.

(i)

(ii)

Plate 2.4   Virus disease symptoms
  (i) Susceptible tomato cultivar with symptoms of tomato mosaic virus (strain O)
  (ii) Tomato fruit symptoms induced by tomato mosaic virus
(Photos: A. Brunt, GCRI)

Mechanical transmission can be affected by chemicals in the donor sap, a feature of certain species only. *Chenopodium* spp and *Dianthus* spp have certain enzymes and polysaccharides present in the leaf sap that inhibit the successful transmission of virus particles. Similarly, the host plant can vary as an efficient receptor and the investigator needs to be aware of these idiosyncrasies as, in many cases, different aspects of study of a particular virus requires different species of host plant for maximum expression of virus effect. The efficiency of transmission can also be affected by pretreating the host plant. Plants kept in the shade, well-watered and at a high temperature of about 36°C will be better receptors than conventionally grown plants.

By far the most frequent method of virus transmission is by vectors which feed on or parasitize the host plant and, of these, the most important are insects of the order *Homoptera* within which the family *Aphidae*, the aphids, are the most common but many leafhoppers (*Cicadellidae*) and whiteflies (*Alleyrodidae*) are also very much involved.

A distinction must now be made between the different methods of virus transmission by aphids. The majority of aphid-transmitted viruses are **non-persistent** or, as defined by some virologists, **stylet-borne**. In this method, the transmission is effected on penetration by the stylet of the host tissue without the virus entering the insect. By contrast, some viruses do enter the insect's digestive system and its blood stream and, after multiplication, can be transmitted during insect feeding. The latter are called **persistent** or **circulative** viruses. A third category, the **semi-persistent** virus is intermediate inasmuch as, whilst they are certainly not circulative, the insects remain infective for several days after acquiring the virus.

As the names imply, there is a difference in the period of infectivity between the groups, with the insect losing the ability to transmit non-persistent viruses within 4 hours compared with the persistent viruses which in some cases retain infectivity for life. There is also a difference in the acquisition feed time which, with non-persistent viruses might only be seconds or minutes compared with a relatively long acquisition feeding time in persistent viruses of perhaps 6–24 hours.

Examples of the three groups are:

(a) *Non-persistent:* cucumber mosaic, lettuce mosaic, potato virus Y, soybean mosaic and sugarcane mosaic.
(b) *Persistent:* barley yellow dwarf, beet western yellows and pea enation mosaic.
(c) *Semi-persistent:* beef yellows and parsnip yellow fleck.

Of the aphid vectors, the peach-potato aphid, *Myzus persicae*, is the most important and is responsible for the transmission of such important diseases as potato leaf roll virus. This insect has a complex life-cycle (Figure 2.1) which involves a primary and secondary host plant. Primary hosts, often trees or shrubs (*Prunus* spp. in this case) are the sites of egg-laying and overwintering whilst secondary hosts, often herbaceous agricultural crops (potato is an example) allows for insect multiplication by asexual reproduction.

Leafhoppers, known to transmit at least thirty different viruses and the most important transmitters of mycoplasma-like organisms, are second to aphids in importance as vectors but whiteflies, a scourge in themselves, are also of significant importance in the spread of virus diseases, especially in the tropics where they cause mosaic, tobacco and cotton leaf curl and many bean virus diseases.

In addition to insects, nematodes and fungi are also involved in virus transmission with nematodes being responsible for the spread of raspberry ringspot virus and tobacco rattle virus whilst fungi such as *Olpidium brassicae* transmit lettuce big vein virus, *Polymyxa graminis* the wheat mosaic virus, and *Spongospora subterranea* the potato moptop virus. Within the nematode-transmitted viruses, a distinction is made on the basis of shape, the rod-shaped particles being members of the **Tobravirus** group, such as tobacco rattle, and the polyhedral particles in the **Nepovirus** group, such as raspberry ringspot virus.

Plant Pathology

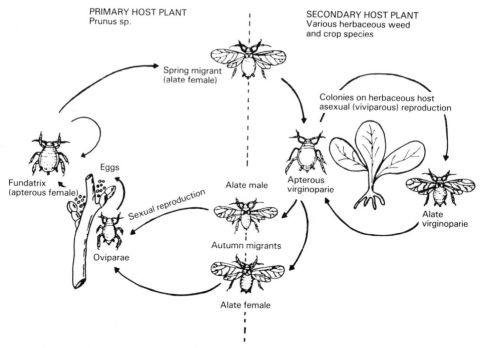

PRIMARY HOST PLANT
Prunus sp.

SECONDARY HOST PLANT
Various herbaceous weed
and crop species

Spring migrant
(alate female)

Colonies on herbaceous host
asexual (viviparous) reproduction

Eggs

Fundatrix
(apterous female)

Alate male

Apterous
virginoparie

Alate
virginoparie

Sexual reproduction

Autumn migrants

Oviparae

Alate female

Fig. 2.1 *Life cycle of* Myzus persicae *(the peach-potato aphid)* (after Walkey, D. G.A., 1985)

Lastly, virus transmission is also accomplished through the agency of seed and pollen, vegetative propagules and grafts and, as is often used for experimental purposes, the parasitic plant species dodder (*Cuscuta* spp.). A number of seed-transmitted viruses are of economic importance and, in all, about forty different viruses are known to be transmitted in this way. Tobacco mosaic virus is an example in which the virus is carried on the testa and cucumber mosaic and soybean mosaic are examples of embryo transmission. Several of the seed-transmitted viruses can also be transmitted in the pollen, alfalfa mosaic and lettuce mosaic being good examples, the virus eventually passing into the ovule through the pollen tube during fertilization. Vegetative propagation is a widespread practice especially in horticulture and it is inevitable that bulbs, corms, tubers and rhizomes can transmit virus from the infected parent plant. Cuttings and runners can also act in this capacity and viruses can also pass along a graft if the graft partner is susceptible. The use of dodder in the experimental transmission of virus was a logical extension of what is known to occur in nature, albeit of very little agricultural importance. The virus does not have to multiply in the dodder for successful transmission, although this does occur and appears to enhance virus movement across the biological bridge.

Nematodes

Plant parasitic nematodes, the *Nematoda*, although causing considerable damage and often proving difficult to control, have traditionally been neglected by the conventional plant pathologist and left to the specialist nematologist. These minute parasites, often referred to as eelworms, are difficult to identify and tedious to extract from plant and soil; consequently they would have not had the research effort devoted to them as their importance would warrant.

The average size of soil and plant nematodes is about 1 mm (1000 μm) and their worm-like shape allows easy passage through the soil and within plant tissue. Not all plant nematodes become **endoparasitic**, some live wholly in the soil, the **ectoparasites**, but both are classified into two groups: (a) migratory and (b) the sedentary. Of the migratory endoparasites, species of *Ditylenchus* causing stem and bulb disorders and *Pratylenchus* spp., the lesion nematodes are well-known examples whilst, of the sedentary endoparasites, the cyst nematodes (*Heterodera* spp.) and root-knot nematodes (*Meloidogyne* spp.) are of world-wide importance (Plate 2.5).

Examples of migratory ectoparasites are *Longidorus* spp., the needle nematodes which have grasses, cereals, vegetables and strawberries among their host-range and feed on young roots by inserting their spear, or stylet, causing galls and distortions. *Trichodorus* spp. are also migratory and feed by sucking out the contents of root hairs and epidermal cells. Winter wheat and brassicas can be badly stunted if infestation is high and sugar beet is especially susceptible.

Paratylenchus spp. furnish examples of the sedentary ectoparasites with the pin nematodes being quite common in the United Kingdom and in Europe.

It is never easy to make an accurate initial diagnosis of nematode attack, as the symptoms are often non-specific such as poor growth, stunting or wilting. However, the common names of many nematode diseases accurately describe the symptom picture—for example, *Heterodera schactii*, the beet cyst-nematode, *Meloidogyne naasi*, the cereal root-knot nematode, *Anguina graminis*, causing leaf galls on many species of grasses, and *Ditylenchus dipsaci*, the stem and bulb nematode which can attack a wide host range including oats, clovers, sugar beet and various bulbs as well as several weed species such as chickweed (*Stellaria media*) and cleavers (*Galium aparine*). Many of the symptoms are caused as reactions to the mechanical damage associated with nematode infection but it is also known that some galls are produced on the induction of plant hormones. In addition, toxins and enzymes are secreted by nematodes which can degrade plant tissue and release the cell contents to sustain nematode nutrition.

Control of nematode diseases is aimed primarily at the use of resistant cultivars, seed treatment, crop rotation and the use of various nematicides. Unfortunately, the breeding programmes for nematode resistance have not been favoured with as high a priority rating as those for resistance to fungal pathogens. Nevertheless, cultivars of oats and barley have been released into commerce with resistance against the cyst and stem and bulb nematodes especially at the Plant Breeding Institute, Cambridge and the Welsh Plant Breeding Station, Aberystwyth, UK. Seed treatment is a standard practice to destroy nematodes which may be adhering to the seed coats. Seeds of lucerne, onion and clover are often fumigated with methyl bromide to control stem nematodes. Crop rotation is the most effective method of controlling cyst-nematodes, and often a four-course rotation is sufficient to reduce the soil population to a safe level. Various nematicides are now available and they may be used as soil fumigants, which usually kill the eggs within nematode cysts, or as drenches of organophosphorous compounds or axime carbamates which normally prevent nematode multiplication.

Mineral Disorders

As we have seen, the growth of crop plants may be adversely affected by a number of biological agencies but there are also several other causes of unthriftiness. In every crop, imbalance of essential nutrients in the soil will result in some form of physiological disorder. Arnon and Stout (1939) have defined the criteria of essentiality as '(i) omission of the element causes failure of growth or reproductive process, (ii) the element cannot be replaced by another in these or all respects, (iii) the element is uniquely associated with an essential metabolite'.

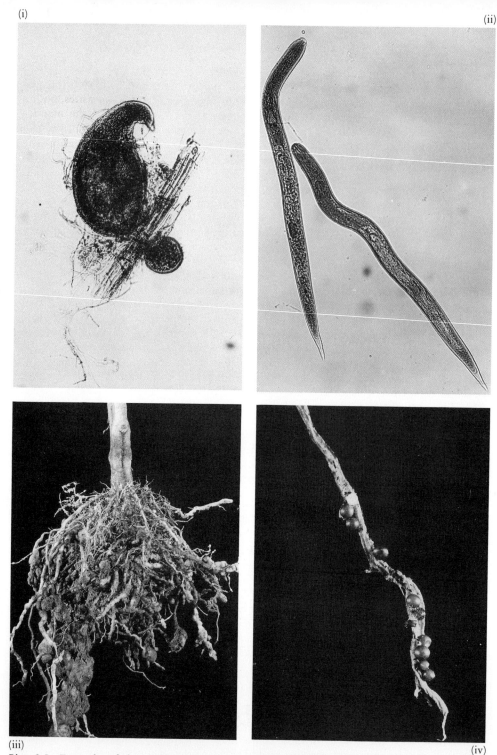

(i)

(ii)

(iii)

(iv)

Plate 2.5   Examples of plant pathogenic nematodes and symptom
  (i) Cabbage root eelworm (*Heterodera cruciferae*) showing 'double' cyst
 (ii) Cabbage root eelworm (*Heterodera cruciferae*) showing 2nd stage larvae
(iii) Root knot eelworm (*Heterodera marioni*) showing tomato roots badly knotted
(iv) Potato root eelworm (*Heterodera rostochiensis*) showing cysts in white and golden stages on
     potato root
(Photos: Shell Photographic Unit)

It may be that the essential nutrient is deficient or in excess, or it may be a case of its presence in an unavailable form due possibly to soil acidity factors or waterlogging. In any event, mineral disorders or imbalances pose considerable problems in diagnosis and many produce symptoms, indistinguishable at some stages, from diseases caused by biological agents.

Nitrogen (N)

Of the major plant nutrients, nitrogen is the most important for most crops and nitrogen deficiency certainly the most common, especially in greenhouse crops. In field crops, nitrogen levels may have been seriously reduced by previous cropping and further losses will depend upon the level of leaching in the drainage water. Nitrogen is essential for protein synthesis, it is present in the nucleic acids DNA and RNA and in many precursors of these vital compounds. Sources of nitrogen are (a) the soil, where the plant can utilize nitrate and various organic forms of nitrogen; (b) biologically fixed nitrogen, either from the free-living nitrogen fixing bacteria, such as *Azotobacter* spp., or from the very productive symbiosis between legumes and nodule-forming bacteria of the genus *Rhizobium*.

Nitrogen deficiency causes a reduction in growth rate accompanied by mild yellowing (chlorosis) of the youngest leaves. Often, especially in cereals, tillering will be sparse and the stems thin and showing reddish tints. In the United States, nitrogen deficiency causes a condition known as yellow berry of wheat in which the grain is starchy, yellow, soft and very low in protein.

Phosphorus (P)

In some plant species, phosphate deficiency is quite easy to diagnose. In tomatoes, for example, the abaxial surface of the leaves can become markedly purple in colour, and there is often a tendency towards dull, bluish green leaves in brassicas, carrots and sweet corn.

In many soils, phosphorus is rapidly fixed to become unavailable. This is the general reaction when inorganic phosphorus is applied to the soil, and in cereals this leads to a slow growth rate, poor tillering as well as the purple tints in the leaves and often the heads. As with nitrogen, this element is essential in the formation of nucleic acids and protein and deficiencies have a pronounced debilitating effect on plant metabolism.

Potassium (K)

Deficiency symptoms are not as common as for nitrogen or phosphorus and are mostly found on lighter sandy and chalk soils. In vegetables, the general symptoms are marginal scorching of the leaves, often preceded by marginal spotting. Severe deficiencies may also produce wilting and early abscission. In cereals, slight deficiencies will produce purplish-brown spots on the leaves, but more severely affected plants will be stunted, weak and producing more tillers than normal although many of these will be infertile.

Magnesium (Mg)

This element is an important constituent of chlorophyll, and ribosomes and many enzymes are activated by this divalent cation. Deficiency symptoms reflect its role in photosynthesis with much interveinal chlorosis of the older leaves in plants such as beans, brassicas and oats, with maize also showing some red and purple tinting. In general, magnesium-deficient cereals are stunted and stiff as well as being somewhat chlorotic. A quick remedy has been the application of soluble magnesium sulphate, Epsom salts for example.

Copper (Cu)

This rarely presents a problem except on peaty or organic chalky soils. Newly reclaimed land can be deficient in copper which is a most important constituent of several enzymes, being tightly bound to the protein. Wheat and maize are particularly sensitive to copper deficiencies, the leaves becoming chlorotic, twisted and often trapped in a loop by the subtending leaf. There may also be an associated failure of the heads to emerge in wheat. Carrots, lettuce and onion are very susceptible, with carrots often suffering yield loss without the appearance of symptoms. In general, copper deficiency results in new leaves being chlorotic and sometimes very bleached in colour.

Iron (Fe)

This element plays an important role in haem synthesis, the activity of haemoglobin in root nodules on legumes and in chlorophyll synthesis and hence photosynthesis. Almost without exception, deficiency symptoms are first detected as chlorosis in young, rapidly expanding leaves. The chlorosis is often interveinal giving rise to a 'tramline' effect in species with parallel views. As the deficiency becomes more serious the chlorosis often becomes almost white with very little trace of green. Purple or pink tints appear in tomato plants deficient in iron and the fruits often ripen to an unusual orange colour.

Manganese (Mn)

The disorder of oats known as grey speck is probably the best known example of manganese deficiency which can occur on alkaline soils with a high organic matter. The symptoms on this host are light-green spots which later turn to grey striping between the leaves. In general, the symptoms resemble iron deficiency but are distinguishable by the characteristic spotting effect which are absent with iron deficiency. Another classical symptom is the 'Marsh spot' of manganese-deficient peas where the seed cotyledons have sunken brown necrotic areas on their inner surface.

Several other elements can cause disorders in plants if in limited supply. Of these zinc, boron, molybdenum, and calcium are probably the most common. For the correction or prevention of mineral deficiencies in UK soils, the reader is referred to *Diagnosis of Mineral Disorders In Plants* (Anon, 1982).

3

The Diagnosis of Plant Disease

Introduction

Diagnosis, or the process of identifying disease, is the first step in the logical, efficient planning of disease management strategies. In the first instance, it is important to determine if a disease is caused by a biotic or an abiotic factor, and implicit in this decision is the ability to recognize that the plant is diseased. An acceptable definition of a diseased plant is 'when its growth, development or appearance deviates significantly from normality'. Mostly, the effects of biotic or infectious agents are well defined and characteristic even to the point that, in some diseases, the presence of the pathogenic organism itself may be apparent. The effects of the many environmental or nutritional deficiencies or extremes often complicate identification especially when a plant pathogenic organism is also active. The correct and rapid diagnosis of a disease problem is of obvious importance, and an essential prerequisite of the advisory pathologist is that, at the very least, he is well versed in this particular branch of his science.

Symptomatology

In any investigation of a diseased plant, the observer must be able to distinguish between **symptoms** and **signs**. A disease **symptom** is the visible manifestation of the plant/pathogen interaction. Symptoms of disease may be localized or extensive and will vary considerably depending upon the nature of the plant species and the invading organism. In many diseases, the pathogen grows or produces various structures on the surface of the host plant. These structures may be in the form of superficial mycelium as in the powdery mildew fungi, sclerotia, runner hyphae or various sporing structures such as the sexual ascocarps of the Ascomycotina or the asexual fruiting bodies such as the pycnidia or acervuli of many of the Deuteromycotina. Wherever such structures are visible they are described as **signs** of the pathogen.

If the pathologist is to become a good diagnostician he must become conversant with all aspects of crop growth, crop production practices and with the range of diseases of the crop in question. He must become well acquainted with any descriptions, photographs or drawings of the diseases that may be available. If he is familiar with the normal growth, development and senescence of a particular crop, abnormalities will become more apparent. Pathologists are often asked to practise their art on badly decomposed or dried-up plant specimens brought in to the laboratory by the curious or anxious grower. However, where possible the disease should be examined *in situ* and, if a sample of plants is to be taken for further observation in the laboratory, this should include whole plants with intact root systems, if this is practical.

Although much information and possibly confirmation of diagnosis may be obtained by consulting specialized textbooks, monographs or possibly coloured slide transparencies, there is much to recommend a standard procedure when confronted with a diagnostic problem. A set routine with good records taken will undoubtedly ensure a more accurate and efficient identification if the problem is ever again encountered.

An example procedure for the recognition of cereal diseases will serve to illustrate the general principles involved in disease diagnosis:

1. Be equipped with a sharp penknife and a hand lens (magnification $\times$ 8–10). Inexpensive field microscopes are also now available. Similarly, a soil pH testing kit is inexpensive and is invaluable to eliminate pH as a factor causing disease.
2. By consulting the farmer, compile a history of previous cropping for the field, cultural practices adopted for the present crop (especially fertilizer, fungicides, insecticides, herbicides applied, etc.). Also, it will be most useful to obtain information on the weather situation. In this way it may be possible to deduce that yellowing symptoms were due perhaps to soil waterlogging or the unavailability of soil nitrogen due to a prolonged dry spell.
3. Make notes on distribution of disease in the field. Is disease uniformly distributed or is it in localized patches and are there weeds or volunteer plants associated with these patches? Is disease most severe near the headlands or in the compressed soil of tractor wheelings? The answers to such questions can provide useful evidence to aid identification.
4. On an individual plant basis, all abnormalities should be noted. Roots should be examined for stunting or discoloration, stems may be cut open to look for fungal mycelium; leaves, stems, glumes etc. should be examined for any sporing structures of the pathogen.
5. On returning to the laboratory, further confirmation may be obtained by looking for any characteristic features, for example, bacterial ooze from cut surfaces, followed by the incubation of diseased tissue in moist chambers to enhance sporulation. Isolation and culturing of the pathogenic organism may also be made and proof of identity sought by the reinoculation of healthy plants as in Koch's postulates.

Koch's Postulates

Proof of pathogenicity caused much controversy until the work of the German, Robert Koch, in 1882 who proposed a set of rules to demonstrate pathogenicity in micro organisms of medical importance. These rules were slightly modified by the renowned Erwin F. Smith in 1905 to satisfy the idiosyncrasies of plant pathogens. The rules, known as Koch's postulates, are as follows:

1. The micro-organism must show constant association with the disease.
2. The micro-organism must be isolated from the diseased host and grown in pure culture.
3. The specific disease must be reproduced when the micro-organism from pure culture is inoculated into the host.
4. The micro-organism must be reisolated from the inoculated diseased host.

It will be immediately apparent that there are limitations to these rules as some pathogens are incapable of reproducing independently of the living host, for example, some biotrophic fungi and plant viruses. However, Koch's postulates are widely used to identify a wide range of fungal and bacterial plant pathogens.

Common Disease Symptoms

There are many types of disease symptoms and, although each might be clearly recognizable, it is important to remember that most pathogens cause more than one symptom and the pathologist is required to form an overall symptom picture for each particular disease. Such a composite pattern of symptoms is known as the disease **syndrome** and is an essential diagnostic feature. Expertise in diagnosis will only come with experience of many disease problems but the expert will be aided by being able to characterize each symptom into a particular group depending on the processes by which they develop. The following are the common disease symptoms.

Chlorosis
This is a yellowing of the tissue which, in some cases, may be due to chlorophyll degradation and, in others, to failure in the formation of this pigment. Chlorosis may be general over much of the plant tissue or it may be locally situated as extensions to necrotic spots, stripes or other disease lesions.

Necrosis
This may be defined as cell death and becomes visible with the death of many cells or areas of plant tissue. Necrosis may be expressed in a number of ways: as **rots** of roots or root systems; as **soft rots** or **dry rots** of fleshy storage organs; as **damping-off** causing the collapse of young seedlings especially under moist, crowded conditions; as **cankers** producing sunken lesions in the stems and branches of woody species; as **anthracnose** producing sunken necrotic areas especially on leaves, pods and other fruit; as **leaf spots**; as **blights** or extensive tissue death with associated browning of leaves and floral organs; or as **scabs** which tend to be local and almost superficial or slightly raised necrotic layers.

Hypoplasia
This is a stunting of plant organs or a general dwarfing of the entire plant, and is a very common symptom in some virus diseases.

Hyperplasia
This is the excessive growth of some plant parts or even the entire plant. It may be due to both the increase in cell numbers (hyperplasia) and in cell size (hypertrophy).

Hypertrophy
This is the distortion of plant parts due to the increase in size of cells. Typical examples are galls and cankers and witch's brooms.

Wilting
When caused by pathogenic organisms, this symptom is progressive and normally irreversible. It can be caused by excessive root damage or death, or by the blocking of the normal movement of water within the vascular tissue of the infected plant.

Transformation of Organs
Many plant pathogens are very specific in terms of the plant organ they primarily attack. The best known examples are the ergot and smut fungi. In the ergot disease of cereals and grasses caused by *Claviceps purpurea,* the whole of the grain is transformed into a large,

compacted mass of mycelium, the sclerotium or ergot (Compendium Plate 3(iv)). Similarly, in loose smut of wheat and barley (*Ustilago nuda*) the whole ear at emergence is transformed into a mass of black teleutospores(Compendium Plate 3(iii)), and in *Ustilago violacea*, the causal organism of anther smut of members of the Caryophyllaceae such as campions, the anthers at maturity are transformed to the extent that they contain masses of black teleutospores but no pollen.

Miscellaneous

This category of disease symptoms embraces many diseases which are of considerable economic importance, for example: the **rusts** with their many small erupting pustules of a brown, orange or yellow colour on stems and leaves; and the **mildews** which attack a wide range of plants and are characterized by producing greyish, superficial mycelium with associated chlorosis and necrosis on leaves, stems and fruit.

Gummosis and Other Characteristic Symptoms

There are many other symptoms which are characteristic of certain diseases, the excessive gum formation or **gummosis** associated with diseases of trees, the sticky **ooze** produced by many bacterial diseases, the reddening of leaves produced by some virus diseases and often with nitrogen deficient plants, stem necrosis may also cause progressive **dieback** of organs and leaf necrosis may also lead to the dead tissue falling away leaving a **shothole** effect. Most plant pathologists will be able to make an initial diagnosis after a preliminary study of the symptoms produced. Further confirmation may be required and it might be necessary to carry out microscopical examination of plant morphological changes or even by the reproduction of the original symptoms upon artificial inoculation of healthy plants (see Koch's postulates).

Field Symptoms

There is undoubtedly a mystique about the identification of diseases in the field. Established pathologists will often jokingly flaunt their expertise to young trainees with instant recognition from the farm gate or by the smell of a crop. It is true to say, however, that there are many patterns of disease distribution, colour changes, associated weed growth and similar abnormalities that give a clear indication to the expert of the identity of the pathogen and of certain facets of epidemiology.

A field of oats susceptible to crown rust (*Puccinia coronata*) will often have an orange appearance, wheat infected with yellow rust a bright yellow appearance and, at certain times, clouds of urediospores of the appropriate colour will be released and easily visible above the crop. The glume blotch pathogen of wheat (*Leptosphaeria nodorum*) produces necrotic lesions on the leaves and glumes which are relatively easy to identify, but more recognizable is the gingery appearance of the distal portion of infected leaves, easily discernible at some distance and due to premature senescence brought about by a fungal toxin.

The distribution pattern of disease can also be very informative. Late blight of potatoes (*Phytophthora infestans*) can either enter a field by the planting of infected tubers or by the spread of spores from a neighbouring field infected earlier or from spores released from discard or cull piles of tubers. The early pattern of disease from an infected seed tuber is a fan-shaped diseased area with the original infection source at its base. Such **primary foci** of infection are readily visible in the early stages of an epidemic and, similarly, disease spreading windwardly away from a corner of the field will usually indicate a cull pile in that situation.

Patchy distribution of disease in a field usually reflects that the pathogen was soil-borne, such patches usually becoming colonized by weeds if plant death occurs sufficiently early. Weedy patches in a cereal field normally indicate the activity of a root or stem base pathogen, the take-all disease of cereals (*Gaeumannomyces graminis*) being a good example.

(i)

(ii)

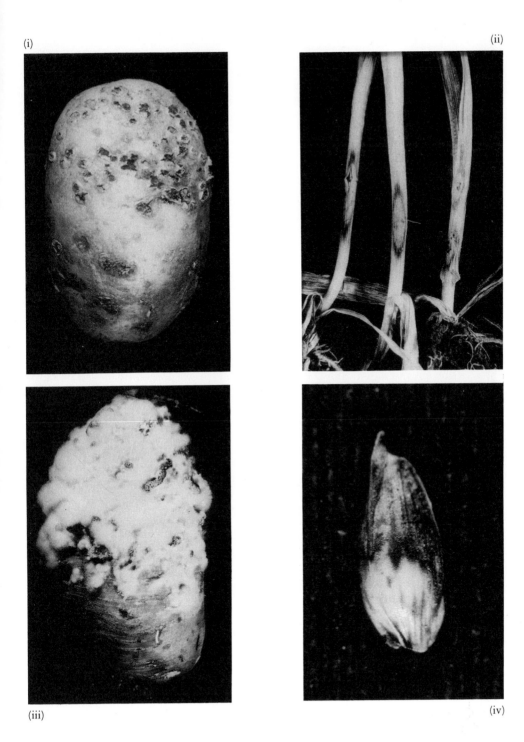

(iii)

(iv)

Plate 3.1 Symptoms and signs of disease
  (i) Powdery scab of potatoes (*Spongospora subterranea*)
  (ii) Eyespot of wheat (*Pseudocercosporella herpotrichoides*)
  (iii) Cottony rot of vegetables (*Sclerotinia sclerotiorum*)
  (iv) Glume blotch of wheat (*Leptosphaeria nodorum*)

It must be emphasized that weedy patches may also have been caused by poor seed drilling, a frost pocket which has killed the emerging crop, or by a waterlogged area where the sown species has died out after suffering 'wet feet'. Uniform distribution, on the other hand, either implies a distant source of inoculum which was air-borne and has settled on the crop, or that the pathogen was seed-borne, the random distribution of seed giving the uniform pattern of disease. A pattern of abnormal growth that often occurs is when a narrow band of unthrifty chlorotic plants is seen across a field. Closer inspection often reveals that the band coincides with the width of the fertilizer drill, the implication being that no fertilizer was applied to the band in question. A most unusual pattern of disease distribution is reported in W. E. Fry's (1982) book *Principles of Plant Disease Management* where he describes a situation in which a single trellis of grapes about 300 m long had been destroyed by some mysterious condition. The consultant pathologist determined that there had been severe electrical storms in the area prior to the vines dying. His diagnosis was that lightning had struck the wire and had been transferred to all the vines which had been attached. Such items of information can be of great value in the ultimate diagnosis.

Individual plant symptoms can be of great diagnostic value to the expert even though it would still be necessary to carry out further tests for positive confirmation. A stunted, red-leaved oat crop is immediately suggestive of barley yellow dwarf virus as a wilting chlorotic brassicae plant is indicative of club root (*Plasmodiophora brassicae*).

Positive identification of the former would require laboratory tests for virus presence, and identification of the latter can be achieved by digging up the suspect plant and examining the root system. The presence of galls alone is still insufficient for identification, but a section cut through a sample of galls indicating grey, mottled tissue will implicate the club root pathogen whereas the presence of a cavity harbouring a small grub eliminates club root and identifies the turnip gall weevil.

Ideally, field recognition should be sufficient but, in an ever-changing situation of crop genotype, environment and pathogens, it is always better to use the field symptoms as an aid to a more accurate, possibly laboratory confirmation, especially when the identification might be the basis for an expensive control treatment.

Classification of Fungi

The taxonomy of plant pathogenic fungi is a subject which is given low priority by pathologists unless identification, at least to the correct genera, is easily achieved. Whilst detailed knowledge of classification criteria is mostly unnecessary, the practising pathologist will undoubtedly benefit from a basic understanding of the systematic arrangement of fungi into families, orders, classes and divisions.

Numerous classification schemes have been proposed, and some have been adopted, but the whole subject is tentative and pathologists are only too aware of the many changes that regularly occur in classification if not in nomenclature as well. A generally accepted scheme for fungal classification has been published by the Commonwealth Mycological Institute (1983) and is reproduced in modified forms in Tables 3.1 and 3.2 with certain items of explanatory information added to facilitate a better understanding of the distinguishing features of the major groups. It will be observed that there are common endings for order, class, division, names etc. These are:

Divisions	—mycota
Subdivisions	—mycotina
Classes	—mycetes
Subclasses	—mycetidae
Orders	—ales
Families	—aceae (families not shown in Tables 3.1 and 3.2)

Table 3.1   A General Purpose Summary of the Classification of the Fungi

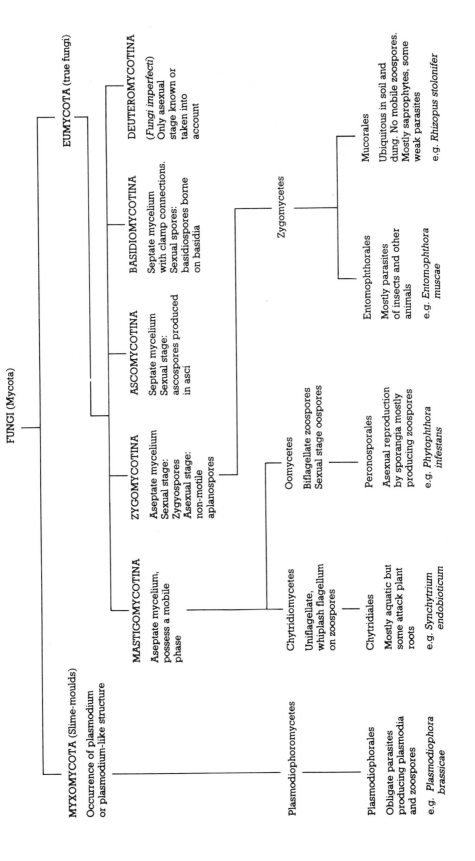

Table 3.2 A General Purpose Classification of the Ascomycotina, Basidiomycotina and Deuteromycotina

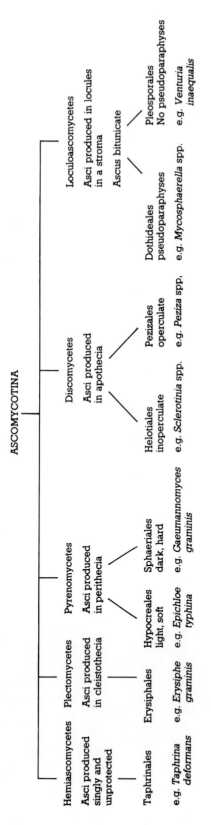

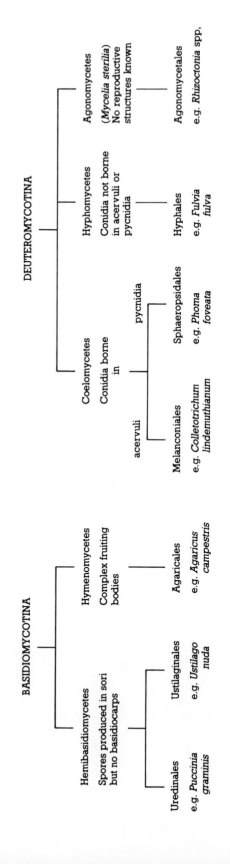

Nomenclature

In any taxonomic scheme, the provision of names for the units or groups is not an arbitrary process decided by the scheme's originator. With the increasing co-operation of national bodies, an internationally agreed Code (International Code of Botanical Nomenclature, ICBN) has been agreed which, although changes frequently occur, does aim to stabilize the general name structure of the Code. Other Codes exist for the bacteria and viruses.

Some of the rules of the Code are obligatory (Principles, Articles) whilst others are optional (Recommendations). The naming of a particular fungal species can be taken as an example. Many names might be given by various scientists to a new fungus. The Code Articles are such that the name to be adopted is the one which was the first to be 'effectively published', that is, in a journal or book and not merely on an herbareum specimen. Having thus decided on the published name, the taxonomist must ensure that this name is applicable to the nominated taxon and if it is, that it does not contravene any Code Article—in other words, that it is 'legitimate' rather than 'illegitimate'. Names other than the correct one belonging to the same taxon are termed **synonyms**. An excellent summary and schematic representation of this procedure is given in *Plant Pathologists Pocket Book* published by the Commonwealth Mycological Institute (1983).

The Code also defines the rules of priority which give the order of precedence to the names of the original or validating author which follow the Latin generic and specific names, for example, *Epicoccum purpurascens* Ehrenb. The Code allows for the differential naming of fungi with more than a single state in their life-cycle (pleomorphic fungi).

One might assume that changes in the names of taxa should not arise unless new knowledge has been obtained. In fact, changes are mostly due to a realization that an incorrect classification has been made after a detailed scrutiny of previously published information. Changes in terms of general placement in the classification scheme are irritating enough, but changes in generic names are the most unpalatable to practising pathologists who may have used the older name for many years. To overcome this problem, a mechanism of retaining well-established names has now been established in the current Code and this 'conservation' of names should certainly give greater stability.

Classification of Plant Pathogenic Bacteria

Although bacteria vary considerably in terms of both morphology and physiology, their classification in classical terms has never merited the same priority as that of fungi. For bacteria in general, there is a very precise classification based on morphology, cultural and biochemical characteristics. This classification is well summarized in *Bergey's Manual of Determinative Bacteriology* (Buchanan and Gibbons, 1974) which, it is probably true to say, is consulted very infrequently by plant pathologists.

Plant pathogenic bacteria are normally identified by the nature of the disease they cause although there are several other tests, including the confirmatory Koch's postulates (see p. 22), which are frequently carried out. As a first step in the identification process, plant pathogenic bacteria can be subjected to the Gram-staining test. This test differentiates bacteria into two groups, those which retain the blue/black colour of the iodine stain, the Gram-positives, and those which lose this stain on washing and take up the red colour of the safranin counter-stain, the Gram-negatives. Only species of *Corynebacterium*, *Nocardia* (one species) and *Streptomyces* give the Gram-positive reaction, whereas the other genera, *Agrobacterium*, *Erwinia*, *Pseudomonas* and *Xanthomonas*, are all Gram-negative.

There is no difficulty in culturing plant pathogenic bacteria, and a simple nutrient agar medium will usually suffice for all species; almost all species will have made good growth after about 48 hours incubation at 25 °C.

Microscopic examination reveals that the bacteria causing plant diseases are all straight or slightly curved rod forms but there is variation in the arrangement of flagella with *Erwinia* species having peritrichous flagella, *Xanthomonas* a single polar flagellum, *Pseudomonas* several polar flagella and *Agrobacterium* subpolar or sparse peritrichous flagella.

The Gram-positive *Corynebacterium* can easily be distinguished on the basis of its peculiar V-shaped or L-shaped (coryneform) cell shapes and up to three polar or subpolar flagella. *Streptomyces* and *Nocardia,* on the other hand, have a most tenuous link with bacteria in general, producing branched mycelium with *Streptomyces* even showing fungal-like characteristics in the form of aerial hyphae with chains of conidia.

Other useful diagnostic features are the fluorescent pigmentation of *Pseudomonas* species on certain media (King's Medium B) or the yellow pigment produced in colonies of *Xanthomonas.* Similarly, some *Erwinia* species can be distinguished by their putrid smell on nutrient medium. Such cultural characteristics can be used for positive identification. The following tests indicate a typical procedure for describing a bacterium.

Tests

1. *Cell morphology:* shape, special structures, flagella arrangement.
2. *Staining reactions:* Gram stain, acid fastness.
3. *Cultural characteristics:* fluorescence, pigmentation, amount and type of growth, temperature and pH tolerance.
4. *Oxygen requirements:* aerobe, anaerobe, facultative anaerobe.
5. *Biochemical characteristics:* action on litmus milk, production of gas both aerobically and anaerobically, production of ammonia, hydrogen sulphide, indole, hydrolysis of starch or gelatin etc.
6. *Phage relationships:* individual species can be identified on the basis of phage sensitivity.
7. *Serological relationships:* agglutination, gel-immune diffusion, ELISA.

Action on the Host Plant

Most plant pathologists would use this character as the most convenient way of classifying plant pathogenic bacteria. There are four main categories of effects of infection on the host plant.

Parenchymal diseases

Many bacterial plant pathogens secrete pectic enzymes which break down the middle lamella of the host cells, a process which results in the maceration of the tissue. The cells die due to water loss and produce various symptoms, such as soft rot, blights, leaf spots etc. (*Erwinia carotovora, Erwinia amylovora, Pseudomonas mors-prunorum*).

Vascular diseases

The main symptom in this category of diseases is wilting caused by a combination of the breakdown products of pectic enzyme degradation of cell wall components and bacterial polysaccharides which eventually block the main transport system, the vessels of the vascular tissue. The invading bacteria, however, may have entered the plant via root, stems or foliage, via wounds or natural openings. A good example is *Erwinia salicis*, the causal organism of Watermark disease of cricket bat willows.

(i)

(ii)

Plate 3.2   Bacterial disease symptoms
    (i)   Crown gall on cherry root (*Agrobacterium tumefaciens*)
   (ii)   Close-up of crown gall on tomato stem (*Artificial inoculation*)
 (iii)   Bacterial die-back on cherry (*Pseudomonas mors-prunorum*)
 (iv)   Fire-blight of pears (*Erwinia amylovora*)

(iv)

Photos: (i) & (iv) East Malling Res. Station; (iii) Shell Photographic Unit)

Systemic diseases

There are very few bacterial pathogens in this category which, in general, implies invasion of all plant parts. Although this definition is too broad for some diseases, by implication they will be seed-borne which is important in the epidemiological sense but offers a good focal point for chemical control measures. Halo blight of dwarf and runner beans (*Pseudomonas medicaginis* f. sp. *phaseolicola*) is a good example with infection of all aerial plant parts from the cotyledons to the leaves, stems and pods.

Meristematic or Hyperplastic Diseases

In this category, the main effect is the stimulation of excessive growth in the form of galls or the proliferation of roots or shoots—the witch's broom effect. The classical example in this category is *Agrobacterium tumefaciens*, the causal organism of Crown gall (Plate 3.2(ii)). In this disease, the pathogen stimulates both cell division (hyperplasia and cell enlargement (hypertrophy).

It must be emphasized that the above categories are very arbitrary and are used very much for convenience. Genera of the same order can produce very different effects and will be grouped in separate symptom categories. Fortunately, only three orders of bacteria incorporate plant pathogenic bacteria:

> Pseudomonadales: *Pseudomonas, Xanthomonas.*
> Eubacteriales: *Agrobacterium, Erwinia, Corynebacterium.*
> Actinomycetales: *Streptomyces.*

4

Infection and Colonization

Introduction

The course of events leading up to the establishment of physiological contact between pathogens and their hosts may be thought to start with the penetration of the host surface, either directly or through some wound or natural opening. In fact, the complete picture also includes the period immediately before penetration when, in many instances, germination of pathogen propagules takes place (Plate 4.1). Infrequently, penetration is affected by hyphal contact after spreading through the soil from a nearby infected host. Such is the case with the tree-rotting fungus *Armillaria mellea* and the cereal root infecting, take-all pathogen *Gaeumannomyces graminis* but the vast majority of diseases of economically important crop plants are infected after the germination of fungal spores or bacterial cells on their surface.

Infection

Pre-penetration

Germination is the starting point of the **pre-penetration** events in the infection cycle. Spores of some parasitic fungi germinate in water alone but many spores, particularly resting spores, must be stimulated to germinate by physical and chemical factors although some parasitic fungi, like the rusts, can germinate without external stimulation from their hosts. Simple experiments can be devised to demonstrate the stimulating activity of substances which diffuse from or through intact cuticles. If drops of water are placed on plant surfaces and left for 1–2 days, it can be shown that the accumulated substances, possibly inorganic ions, can have highly stimulating effects on the germination of certain pathogen spores; chemicals from rose petals activate spores of *Botrytis cinerea,* for example.

There is a wide variety of chemicals with known stimulatory properties on spores *in vitro*. In the rose, it is sugars such as sucrose, glucose and fructose which are responsible but wetting agents, by increasing the availability of water to the spore protoplasts, also

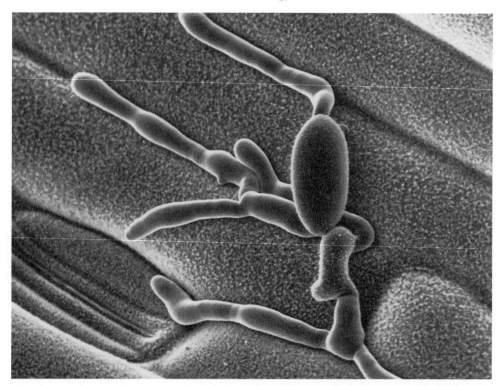

Plate 4.1 Spore germination—a conidium of wheat powdery mildew (*Erysiphe graminis*) germinating on a leaf surface. Note stomate bottom left. (Photo: ICI Plant Protection Ltd)

produce stimulatory effects as do ethanol and acetone and other organic solvents such as chloroform and ether although, in the latter two substances, it is their fat solvent ability that promotes germination rather than their effect on surface tension.

Root exudates are particularly significant in the pre-penetration activities of many soil fungi. The zoospores of certain *Phytophthora* species are chemotactically attracted to their hosts' roots where they encyst prior to entry. Roots also exert an indirect effect on the germination of pathogen spores in that area immediately surrounding them (the **rhizosphere**). This is done by stimulating the growth of certain rhizophere organisms, many of which are either antagonistic to potential pathogens or out-compete them for the available nutrients, especially the large amounts of organic substances released into the rhizosphere by the roots.

Competition is also a factor in the success of pathogens on the leaf surface (the **phylloplane**) but to a lesser extent than in the soil. In addition, the plant surface can also exude substances which are inhibitory from orange peel and coloured skin onions, the latter producing protocatechuic acid and catechol which strongly inhibit the germination of the onion smudge fungus, *Colletotrichum circinans*.

The germination stage is exemplified by either the production of a short hypha, the **germ-tube**, or by the release of zoospores. In the case of *Phytophthora infestans*, the sporangium has the capacity to do both depending on temperature; below 15 °C zoospores will be released, and above 15 °C the sporangium behaves like a conventional conidium and produces a germ-tube.

Penetration

Once germination has been accomplished, the next stage in the infection cycle is that of **penetration**. Three routes of entry are available for pathogens: wounds, natural openings or directly through the intact surface of the host.

Wound Parasites

It is a most unusual fact that many diseases are caused by pathogens which do not possess the facility of directly penetrating through the intact plant surface. These wound parasites are most important factors in the economics of growing potatoes, fruit and many root crops and vegetables. Wounds may be caused by man and his machines as well as by physical agencies such as wind and hail and the many and varied insect and other animal pests. It has already been emphasized that the entry of viruses into their hosts generally involves a secondary organism, the **vector** (see Chapter 2).

For most of the growing season, plants have very successful natural barriers to infection. These may be of an active or a passive nature (see Chapter 6), but occasionally these barriers may be breached. High winds in early autumn can defoliate trees before the natural abscission layers form. The apple canker fungus, *Nectria galligena* and the cherry canker bacterium, *Pseudomonas mors-prunorum* often effect entry through such wounded areas. The emergence of lateral roots not only open up the cortical cells to infection, as with the tobacco root fungus, *Thielaviopsis basicola*, but also such disruption can result in the release of substances rich in carbohydrates and amino acids which attract zoosporic fungi such as *Phytophthora* spp.

Wounds need not be large, in fact lesions produced by other pathogens can produce sufficient wounding to allow the entry of secondary invaders. The cracking of scabs on apples and pears due to *Venturia inaequalis* is a common entry point for the brown rot fungus, *Sclerotinia fructigena*, and potato blight lesions are often colonized by the bacterial soft-rot organism, *Erwinia carotovora*, with disastrous consequences.

Wounds produced by insects and other animals can vary from the minute punctures of stem-borers to the extensive defoliation or root or stem damage caused by slugs. Austwick (1958) has estimated that some forty-five fungal diseases of higher plants are transmitted by insects. The Dutch elm fungus, *Ceratocystis ulmi*, is introduced into sap wood by its vector, the bark beetle, *Scolytus scolytus*, which not only effects transmission of the fungal spores but creates the entry point by the excavation of its breeding galleries.

Manmade wounds may be deliberate, as in the case of grafting or pruning, or they may be accidental as in many agricultural and horticultural practices which can cause mechanical injury. The harvesting of potatoes with the spinner operating at the wrong depth and at too high a speed produces wounds which allow the entry of numerous storage pathogens including *Pythium ultimum*, *Phoma foveata* and *Fusarium caeruleum*. Many post-harvest fruit diseases are caused by the entry of pathogens by wounding on the tree or during picking and harvesting. Minute wounds caused whilst the fruit are still on the tree allow the entry of *Penicillium expansum*, *P. digitatum* and *P. italicum* as well as *Gloeosporium fructigenum* and *Sclerotinia fructigena*. Careful picking procedures to keep the stalk intact on apples and pears, and techniques to minimize bruising at all times during sorting and transport, will result in significant reductions in damage by these wound parasites.

Several agricultural and horticultural practices have also been shown to facilitate the entry of wound parasites. The use of wire ties to secure fruit trees to their stakes can cause wounds which allow *Nectria galligena* to enter its host, and the innovative layering technique recently introduced for apples and pears has resulted in the fruit forming only a few centimetres from the soil surface. Such a practice has dramatically increased the incidence of *Phytophthora syringae* which can easily contaminate the fruit by either rain-induced or tractor-induced splashing. A final comment about wounding as a means of entry for plant pathogens is that except in exceptional cases such as the entry of larch canker, *Lachnellula willkommii*, which appears to be well adapted to entry through cracks in the bark induced by frost, there seems to be no difference in terms of pathology whether the wound is self-inflicted or due to external agencies; to the pathogen a wound is a wound.

Natural Openings

Entry through natural openings is very common with bacterial plant pathogens, the entry point frequently being the **stomata.** Water is usually required to facilitate entry, the bacteria present in drops of moisture at the stomatal aperture being drawn into the cavity beneath as the droplet evaporates. The fireblight pathogen of pears, *Erwinia amylovora*, can accomplish entry in this way.

Many obligate parasites enter through stomata. When a rust uredospore germinates on a cereal leaf, it produces a germ-tube which elongates until it reaches a stoma. The end of the germ-tube then swells to form an appressorium which positions itself over the stomatal aperture. Narrow hyphae then pass through and swell to form substomatal vesicles, from which the eventual colonizing internal mycelium develops. The germ-tube is obviously chemically attracted to the stomata, but Gäumann (1950) cites the interesting case of *Botrytis cinerea* which enters through the stomata in diseased *Vicia faba* plants but through the cuticle in healthy plants.

Flexibility of entry mechanism is also possessed by *Phytophthora infestans* which can also enter either through the stomata or directly through the cuticle and epidermis. *Phytophthora* spp., as well as other downy mildew pathogens including *Pseudoperonospora humuli* on hops and *Plasmopora viticola* on vines, can produce mobile zoospores on the germination of their sporangia. On finding an aperture, the zoospores encyst and later germinate to penetrate into the stomatal cavity.

A common point of entry for some pathogens are the **lenticels.** *Streptomyces scabies*, the causal organism of common scab of potatoes, enters in this way as does *Sclerotinia fructicola* which causes brown rot of stone fruits. **Nectaries** afford entry to several bacteria, *Erwinia amylovora* entering its pear host in this way and multiplying rapidly in this highly suitable environment, well supplied with rich sugary compounds. Water pores, or **hydathodes,** facilitate one method of entry for the bacterium *Xanthomonas campestris* into its cruciferous host. Here, the bacteria in water droplets at the edge of leaves are sucked into these natural openings by the plant under drying conditions.

Not to be confused with natural openings is the infection of floral organs by many cereal pathogens. The stigma surface is the entry point for the loose smut fungus, *Ustilago nuda*, although it can also penetrate the pericarp wall at flowering time. In similar manner, ascospores of the ergot fungus, *Claviceps purpurea*, are discharged to coincide with anthesis. They reach the stigma surface either by rain-splash or by insect dispersal. However, it is emphasized that, with both these examples, ultimate infection is by direct penetration albeit within the confines of the floret.

Direct Penetration

The third and most complex method of entry into the host plant is by **direct penetration.** To do this, it will be necessary for the pathogen to either exert considerable physical force or to fashion an entry point by the degradation of the host surface. To an extent, both methods are probably involved. The thick layer of bark on woody plants presents an almost impenetrable barrier to most fungi. Exceptions such as the root-rotting *Armillaria mellea* have developed the necessary mechanical and chemical abilities to allow its rhizomorphs to effect entry.

To fully understand direct penetration, it is necessary to appreciate the physical and chemical nature of plant surfaces. Figure 4.1 is a diagrammatic representation of a section

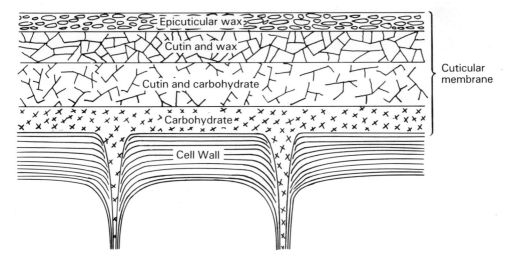

Fig. 4.1 *Diagrammatic representation of the outer layers of an herbaceous plant*

through the external layers bounding herbaceous plants. In practice, there are considerable differences in the structure and composition of the outer layers of various plant species, even between different parts of the same plant. The outermost layer, the **cuticle**, is a spongy framework of cutin interspersed with wax platelets, some carbohydrates, phenols, pectins and fats. The cuticle presents the first physical obstacle to foliar and stem pathogens, roots being protected by a layer containing mainly suberin with additional cutin and lipid materials. The cuticle, being hydrophilic, can swell depending upon the water content of the underlying tissue, its resistance to fungal penetration being inversely related to the degree of swelling.

Enzymes and Penetration

Pathogens participating in direct penetration will normally produce a complex of enzymes and hyphal modifications known as infection structures. In many cases, the fungus hypha attaches itself to the cuticle by means of a swelling, the **appressorium**. From the centre of the lower appressorium surface, a slender **infection peg** grows out and, with some physical pressure and probably considerable enzymatic 'softening up', penetrates the plant surface. The increased surface area of the swelling but anchored appressorium can exert strong downward pressure on the developing infection peg, producing a 'drawing pin' effect. The involvement of enzymes in this process was always conjectured but scanning electron microscopy has produced evidence supporting both the physical force method of entry, where the plant surface is pushed inwards and the entry point ragged, and the enzymatic degradation method, where the entry point is a clean-sided aperture with no indentation. (Plate 4.2). Consensus opinion is that probably both methods operate in the vast majority of pathogen infections. The formation of appressoria does not necessarily indicate direct penetration. Many rust species enter via the stomata after forming appressoria over the stomatal aperture.

The enzyme responsible for the degradation of cutin has been termed **cutinase**, the demonstration that fungi could grow on cutin as the sole source of carbon being taken as

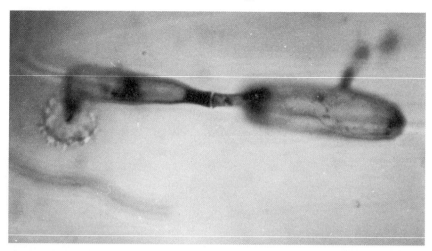

Plate 4.2 Enzymatic degradation of cell wall beneath infection site of *Erysiphe graminis*. Note circular area of enzymatic activity. (Photo: T. L. Carver, WPBS)

evidence that cutin-degrading enzymes were being produced. Cutinase has been purified from the extracellular fluid of cutin-grown *Fusarium solani* f. sp. *pisi* (Purdy and Kolattukudy, 1976), and has since been confirmed as a product of a great number and variety of plant pathogens. Cutinase has a molecular weight of about 25 000 and is a glycoprotein containing about 3–6% carbohydrates.

Confirmation of the involvement of cutinase in cuticular penetration has been provided using antibodies prepared against the enzyme. Again, using *F. solani* f. sp. *pisi* and peas, it was possible to show the production of cutinase at the infection site by staining with ferritin-conjugated rabbit antibodies specific for cutinase (Shayk, Soliday and Kolattukudy, 1977).

Cutinase is produced under the genetical control of the fungus but the gene responsible is only 'switched on' when the pathogen senses it is on a plant surface. Gene activation normally occurs within 15 minutes of the fungal spore being exposed to cutin. The proposed mechanism is that small amounts of cutinase present in the spore initiate the cutin-degradation process, releasing monomers which diffuse back to the spore, triggering the cutinase gene and initiating the production of quantities of cutinase sufficient to facilitate penetration (Figure 4.2). The whole subject of cuticle penetration has recently been reviewed (Kolattukudy, 1985).

Once through the cuticle, the pathogen encounters the cell wall. Cell walls can either be of a primary nature, consisting mainly of polysaccharides, or secondary when they supplement the primary wall with more polysaccharides and, in certain cells, lignin (for example, xylem) or suberin (for example, endodermis).

Most but not all plant pathogens can produce extracellular, cell wall degrading enzymes (CWDE). However, this ability is not necessarily proof of their involvement in pathogenicity. As with the penetration of the cuticle, a knowledge of cell-wall structure will be most helpful if the complexities of its enzymatic degradation is to be understood.

One important aspect of cell-wall structure is that it is of a constantly changing nature as growth and development occurs. In simple terms, the cell wall is composed of layers of microfibrils embedded within a complex continuous matrix. The main constituents are pectic substances, cellulose and hemicellulose with lignin and suberin being the strengthening materials of some specialized cell walls such as the xylem, sclerenchyma and bark cells. Pectic polymers are predominant in but not exclusive to the intercellular region,

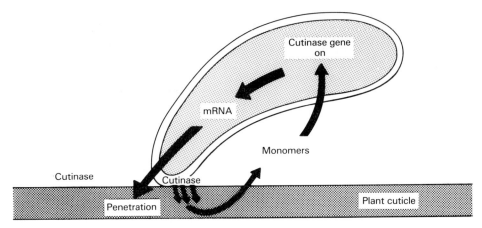

Fig. 4.2 *Schematic representation of how the plant cuticle induces cutinase in a fungal spore* (after Kolattukudy, 1985)

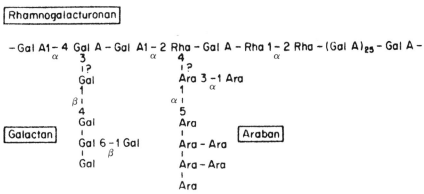

Fig. 4.3 *Possible structure of 'pectic' polymers—rhamnogalacturonan and covalently linked neutral polysaccharides. Gal A, galacturonic acid; Rha, rhamnose; Gal, galactose; Ara, arabinose* (after Cooper, 1984)

or **middle lamellae**. They consist of mixtures of neutral arabinogalactans and acidic galacturonorhamnans. The most important of these pectic polymers, rhamnogalacturonan, comprises chains of α-1,4-linked galacturonic acid residues interspersed with 1,2-linked rhamnose. To this backbone are linked araban (α-1,3- and α-1,5-highly branched) and galactan (β-1,4-, linear) side-chains (Figure 4.3). The pectin polysaccharides are probably present partly as gels and partly as viscous solutions, the latter especially in young cell walls. Pectic polymers constitute a substantial proportion (around 35%) of primary dicotyledenous walls but generally much less (8–9%) in monocotyledons.

The cellulose component comprises a combination of microfibrils and macrofibrils. Each microfibril is an extremely long chain (8000–12 000 units) of β-1,4-glucose residues. Bundles of microfibrils are bonded together to form the macrofibrils, this structure possessing considerable physical strength in a similar but not strictly structurally analogous manner to that of the interweaving wire in the structure of wire ropes. Hemicellulose is also a principal component of primary cell walls in dicotyledonous plants. The main hemicellulose is xyloglucan, a polymer with a cellulose-like backbone of β-1,4-glucose with

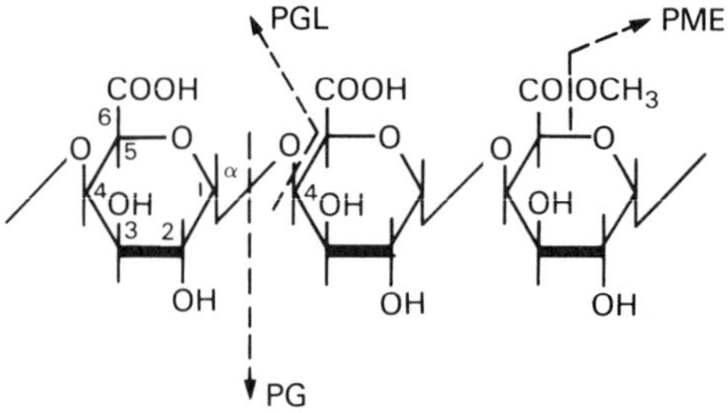

Fig. 4.4 *Cleavage of an α-1,4-linked polygalacturonide chain by* endo-*polygalacturonase (PG) and* endo-*polygalacturonide lyase (PGL), and demethoxylation by pectin methylesterase (PME) (after Cooper, 1984)*

terminal branches of 1,6-xylose and fucose and capable of hydrogen bonding to cellulose microfibrils.

It would seem obvious, bearing in mind this cell-wall complexity, that there will be a requirement for a variety of enzymes if this barrier is to be penetrated. The enzymes responsible can be grouped into two broad categories, the **pectinases** and the **cellulases**.

The first CWDE produced *in vitro* when pathogens such as *Verticillium albo-atrum* or *Colletotrichum lindemuthianum* are grown on extracted cell walls are the **endo-polygalacturonidases** or pectic enzymes. Degradation of primary walls is often extensive and rapid during infection by facultative parasites which cause necrosis and most evidence implicates the pectic enzymes as major factors in this process. Two types of pectic enzymes are responsible for the degradation of rhamnogalacturonan, the hydrolytic poly-galacturonases (PG) and the pectic lyases (PL). Both are chain-splitting enzymes which can either act by splitting the 1–4 glycosidic bond between adjacent uronic acid units by a hydrolytic mechanism (PG) or by a transeliminase system (PL). Both enzymes can either attack at random along the chains (*endo-*) or from the ends of the chain (*exo-*) (Figure 4.4). Pathogens can produce PG, PL or both.

During infection, the sequence of events appears to be the production of pectic enzymes, followed by hemicellulases and then the cellulases. The main hemicellulose, xyloglucan, can be degraded by endo-β-1,4-glucanases, the smaller degradation products being unable to bind to cellulose and hence causing the initiation of structural breakdown. Degradation of cellulose demands a complex of enzymes and the precise process is not fully understood. Although still somewhat speculative, it is considered that one group, the C_1 enzymes, carry out the initial bond breaking which loosens up the bundles of macrofibrils releasing single chains. A second group, the C_2 enzymes, are thought to continue the degradation process, breaking the cellulose fibrils into shorter lengths. The third group, the C_X enzymes, then begin hydrolysing the short chains to produce cellobiose; the fourth enzyme, cellobiase, completes the process by hydrolysing the cellobiose to glucose. A more recent view is that the C_1 enzymes might simply be involved in preventing the reformation of glucosidic linkages (Wood and McCrae, 1979).

Colonization

Following penetration, the extent to which pathogens colonize their hosts and/or cause damage depends upon the nature of both host and pathogen and their interaction with the environment. Host resistance or adverse environmental conditions can severely limit subsequent **post-penetration** events, and the severity of disease is normally closely related to the extent of pathogen invasion.

Extreme differences in colonization occur. With the powdery mildew group of pathogens, colony growth is superficial (that is, **ectotrophic**) with only the occasional epidermal cell being invaded to allow haustorial development. Conversely, some pathogens produce systemic infections, spreading throughout the host plant. Only *Ustilago nuda*, causing loose smut of barley, is truly systemic but other pathogens can also be considered systemic in their affects. The potato late blight pathogen, *Phytophthora infestans*, can start the disease cycle as a seed-tuber infection. The fungus invades the leaves and, when external conditions are conducive, sporulates at the margin of necrotic lesions which rapidly spread, killing off the whole haulm. The soft-rot causing bacterium, *Erwinia amylovora*, once established in the host tissue, will also spread rapidly through the intercellular spaces of the parenchyma tissue producing its characteristic cell-macerating effect through its production of pectolytic enzymes. Systemic colonization is also a characteristic of plant virus diseases where almost all the cells of the plant can become infected especially with the help of the phloem transport of virus particles.

Cell wall damage frequently occurs well in advance of pathogens like *E. carotovora*, the degradation of the middle lamella often being accompanied by the disruption of the host cytoplasm. This is the typical sequence with necrotrophic pathogens, the death of the host cell serving both to provide nutrients for the advancing pathogen and to eliminate any host defensive system.

The relationship between cell death and disease damage is fairly obvious although the timing of necrosis will have a profound effect on yield; for instance, a late infection of wheat leaves by a rust pathogen is likely to be of minor importance compared with similar levels of infection resulting from an earlier initiation of disease. Obligate pathogens like rusts are also of interest in that they produce very little damage until the sporing pustules are produced. The sequence of events after penetration is such that the developing mycelium is restricted to a relatively small colony within the photosynthetic tissue. The uredial pustule develops from this subepidermal mycelial mat and erupts through the host surface. During the 8 or 9 days latent period, there is no great breakdown of host tissue. Visible symptoms such as chlorosis only appear after sporulation, and the plant only becomes severely damaged if there are many individual pustules on the same leaf, not an uncommon situation with such a high sporing pathogen.

Colonization can be highly specific to certain host tissues. Vascular wilt fungi such as *Verticillium albo-atrum* or *Fusarium oxysporum*, will normally be confined to the vascular tissue of the infected host although the fungus will be present in the root cortical zone where entry will have occurred. Once through the endodermis, a resistance barrier in some plants, the pathogen colonizes its host very rapidly by the transportation of microconidia upwards in the xylem.

Toxins

With cell death and consequent loss of photosynthetic tissue due to CWDE, damage due to plant disease can be considerable. Many pathogens can also produce host damage through the action of **phytotoxins**. Phytotoxins can be defined as substances produced by micro-organisms which are injurious to plants, often in minute amounts. Some definitions of toxins would include enzymes but the following discussion will exclude enzymatic activity as this topic has already been discussed.

In chemical terms, toxins are very varied and include peptides, glycoproteins, polysaccharides, organic and fatty acids. The idea that pathogenesis might be due to the production of poisonous substances by the pathogen is by no means new. However, the fact that a pathogen produces a toxin *in vitro* does not necessarily mean that it will also produce it in the host plant; the reverse is also true. Proof of the presence and activity of toxins is now accumulating and involves the demonstration that sterile filtrates from a pure culture of the pathogen can reproduce all or some of the disease symptoms. The purification of the toxin will also allow more precise evidence of involvement.

Toxins need only be produced in minute amounts to cause cell damage and the fact that they are mobile within the plant explains why symptoms may be produced at some distance from the site of infection or from the leading edge of the colonizing mycelium.

Tentoxin

A common symptom due to toxin activity is a general chlorosis. *Alternaria tenuis* produces a toxin which has been termed tentoxin, its effects being a variegated chlorosis of seedlings caused by the inhibition of the development of chloroplasts from protoplastids. The method of toxic action appears to be a blocking of ATP synthesis which would prevent the formation of essential chloroplast proteins.

Tabtoxin

Another toxin, tabtoxin, is produced by the tobacco wildfire bacterium, *Pseudomonas syringae* pv. *tabaci*. The toxin is lactonethreonine:

$$NH_2 - CH - \overset{\overset{\textstyle O}{\|}}{C} - NH - CH - COOH$$

with side chains:

$$\begin{array}{cc} CH_2 & CHOH \\ CH_2 & CH_3 \\ \end{array}$$

$$\overset{O}{\diagdown}\ C - C \\ | \quad | \\ HN - CH_2$$

It is another chlorosis-inducing toxin, the effect being due to the indirect effects of the inhibition of the enzyme glutamine synthetase, a key enzyme in nitrogen metabolism. Chlorosis probably occurs as a result of this metabolic disruption, the necessary protein-N being replaced by ammonia-N. It is interesting to note that tabtoxin is a non-specific toxin, producing chlorosis on both susceptible and resistant cultivars if applied artificially.

T-toxin and Victorin

Several toxins are known to cause necrosis. Race T of *Helminthosporium maydis*, produces a highly specific toxin, the T-toxin which only produces necrosis on maize cultivars carrying the Texas cytoplasmic male sterility factor. The mode of action of T-toxin is not fully understood but it is thought to involve several sites of action. Similar host specificity has been demonstrated for victorin, a toxin produced by *Helminthosporium victoriae*, the causal organism of a seedling blight of the oat cultivar Victoria or any cultivar derived from that parent and containing the single gene conferring susceptibility. Victorin is toxic at concentrations as low as 0.0002 g/ml. Again, the mode of action is none too clear although the disruption of the plasma membrane is a distinct possibility. It is known that, in susceptible cultivars, there is an increase in respiration rate proportional to the concentration of the toxin but this may be an indirect effect of the damage to and increased permeability of the plasma membrane.

Amylovorin and Fusicoccum

Wilting is also a symptom that can be caused by certain toxins. In some bacteria, *Xanthomonas phaseoli* for example, toxic polysaccharide slime is produced, the pathogenicity of the bacteria being correlated with the amounts of slime produced which can physically block the vascular tissue. A similar effect is produced by amylovorin, the toxin produced by the fire-blight of pears bacterium, *Erwinia amylovora*. Wilting can also result as an indirect effect of the toxin on the plasma membrane which makes it freely permeable and hence allows water loss from the cells. Fusicoccum, a diterpene glucoside produced by *Fusicoccum amygdali*, causes a wilting of almond and peach trees in this way.

There can be no doubting the involvement of toxins in many plant diseases. Several less obvious disease symptoms may also be the result of toxin activity but, until toxins can be readily purified and identified, our knowledge of the extent and importance of toxins in disease damage will be seriously limited. The reader is recommended to a review of the role of toxins in plant disease by Turner (1984).

5

Physiological Responses of Plants to Pathogens

Introduction

Many plant symptoms are highly specific to a particular host–pathogen combination. For example, there is nothing else to match the cocoa swollen shoot virus symptoms nor the ergot symptoms of *Claviceps purpurea* on cereals. However, many symptoms are common to many plant diseases, chlorosis, necrosis, wilting for example, suggesting that many pathogens produce similar physiological changes in their hosts. It is tempting to suggest that these changes indicate a common pathway or sequence of biochemical events. The evidence for this has not been forthcoming and, against this hypothesis, such changes can be induced by non-microbial agents. The explanation might emerge when more information is obtained from investigations at the molecular level. In the meantime, this chapter will consider the nature and extent of physiological responses in infected host plants.

We have already seen that necrotrophic pathogens produce almost instantaneous and drastic changes once they become established in susceptible host tissue. Cell and tissue death is often extremely rapid—for example, the killing of haulms in late blight of potato—and, in many cases, the metabolism of the remainder of the plant is only affected to the extent that the dead parts cease to contribute their particular share to the totality of the plant's growth. These deductions may be very important in terms of economic yield, the actual amounts varying depending upon plant growth stage and rate of disease development.

In contradistinction, biotrophic pathogens will colonize their hosts without such immediate drastic changes, interacting at all times with host metabolism and producing changes in respiration and photosynthesis. With the physiology of the diseased plant altered in this way, allied and consequential effects can often occur, nutrient transport and ultimately plant growth being the most obvious.

Respiration

One of the best documented changes in the physiology of the diseased plant is that of an increased rate of respiration. This increase has been observed for diseases involving fungi, bacteria and viruses although the most critical information has been obtained from plants infected with biotrophic fungi. In plants infected with rusts or powdery mildews, for example, respiration rate increases of up to 100% have been recorded by the time sporulation occurs.

Respiration is the complex process by which living organisms obtain energy through the oxidation of organic material to more simple compounds. Two metabolic pathways are involved in the degradation of carbohydrates such as starch and sugars to smaller carbon compounds, the glycolytic Embden–Meyerhof—Parnas (EMP), and the pentose phosphate pathways. Starting with glucose, pyruvic acid is one of the main breakdown products and this is eventually oxidized in the citric acid cycle (TCA) to carbon dioxide and water. Changes in both EMP and TCA can be observed in diseased plants.

The earliest evidence of changes in respiration was provided by Allen and Goddard (1938) with wheat infected with powdery mildew. *Erysiphe graminis* produces superficial mycelium which only has internal contact with its host by the insertion of haustoria into the epidermal cells. Experiments where the mycelium is peeled off indicate that increases in respiration have been triggered off and are maintained even after the removal of the pathogen.

Several possible mechanisms have been proposed for the changes in respiration. The first of these suggests an uncoupling of oxidative phosphorylation. The availability of adenosine diphosphate (ADP) is crucial to the regulation of cell respiration. In healthy plants, the ADP level is kept at a relatively low level under anaerobic conditions by the conversion of ADP to ATP (adenosine triphosphate) in the cytochrome system. It is known that increases in respiration rates can be brought about experimentally when ATP formation is 'uncoupled' from electron flow in the cytochrome system. With the synthesis of ATP reduced or prevented, the concentration of ADP increases, the rate of oxygen increases and hence respiration rates rise. The suggestion is that pathogens can mimic the uncoupling ability of such agents as 2,4-dinitrophenol (DNP), and there is circumstantial evidence inasmuch as the level of ATP is often lower in infected tissues. The uncoupling hypothesis has received support from studies of barley infected with the net blotch pathogen, *Pyrenophora teres* (Smedegård-Peterson, 1980), but it has been challenged by evidence that victorin (see Chapter 4) and other host-selective toxins caused not only increased respiration of infected tissues but also a rapid loss of ions prior to the respiratory rise (Black and Wheeler, 1966). Also, the uncoupling hypothesis has not been confirmed by evidence from biotrophic pathogens, Daly and Sayre (1957) finding that safflower plants infected with the rust fungus *Puccinia carthami* had an increased respiration rate due to the demands of the pathogen for ATP rather than to uncoupling. Biotrophic fungi obviously create a metabolic sink which, at the expense of host energy, draws photoassimilates and other nutrients from surrounding tissues to the sites of infection.

Glycolysis represents one of the major metabolic pathways for the degradation of glucose to pyruvate. Glycolysis via the Embden–Meyerhof pathway can be substituted by an alternative route for the production of pyruvate, the pentose phosphate pathway. Several lines of evidence have indicated a shift fom the glycolytic to the pentose phosphate pathway. The earliest indication came from Daly and Sayre (1957) working with rusted safflower who found clear evidence for the operation of a non-glycolytic pathway. Similarly, Scott (1965) working with barley infected with *Erysiphe graminis* f. sp. *hordei* found only slight increases in glycolytic enzyme activity but a two to three-fold increase in enzymes of the phosphate pathway.

Such changes in respiratory patterns may simply reflect the fact that the pentose phosphate pathway operates at a higher level in fungi than in plants, and it does seem clear that it is only its quantitative importance and the extent to which it operates that requires elucidation.

Evidence also exists of alternative terminal oxidation systems operating in infected tissues. Systems involving phenol oxidases and ascorbic acid oxidases are known to be activated in diseased tissue in addition to the normal cytochrome system. The production of phenolic compounds as a host reaction to infection would certainly involve phenol oxidases but, other than their effects upon respiration, the significance of these alternative oxidation systems is still not clear.

There can be no doubt that infection causes a general stimulation of host metabolism with an accompanying increase in respiration. The most obvious explanation for this is that the pathogen, in causing a drain on host nutrients, stimulates synthetic metabolism. This is most clearly seen in studies of biotrophic fungi which induce the maximum increase in host respiration coincidental with the onset of sporulation.

An interesting controversy surrounds the topic of respiration in resistant plants. It might be assumed that many plant defence mechanism would be intensely energy-dependant. In fact, Daly (1976) suggested that a minimum of 4–5 molecules of glucose would be required to synthesize one molecule of the **phytoalexin** rishitin in potato plants. The evidence to confirm this hypothesis has not emerged, although Scott and Smillie (1966) showed a slight increase in respiration compared with controls using near isogenic lines of barley infected with *Erysiphe graminis* f. sp. *hordei*. However, Bushnell (1970) found a slower increase in respiration in resistant than in susceptible cultivars of wheat infected with *Puccinia graminis* f. sp. *tritici* which was contrary to the findings of Heitfuss (1965). The role of respiration and energy generation in diseased and disease-resistant plants has been reviewed recently (Smedegård-Peterson, 1984).

Photosynthesis

The most common effect of infection of the aerial parts of green plants is a reduction in photosynthetic tissue which, in turn, leads to an adverse effect on plant growth, possibly on economic yield. Sometimes, there may be an effect on photosynthesis that does not manifest itself as visible chlorosis or necrosis. Unlike most foliar infections, there may be no apparent loss in photosynthetic tissue. The mechanisms of plant physiological processes and pathogen attack are now fairly well understood but the means by which these processes are controlled by the plant and modified by the presence of the pathogen are still vague.

A simple model relating infection to reduction in photosynthetic processes might only take into account the area visibly affected, but it has been shown that broad bean plants infected to the extent of 30–40% leaf loss had similar relative growth rates to those of uninfected plants. The implication is that, in adjacent uninfected areas of the leaf, photosynthetic efficiency has been enhanced to compensate for the loss in the infected areas, the undamaged tissue responding to the destruction of the pathogen itself or its metabolites.

Much of the early work on the photosynthesis of diseased plants was concerned with the breakdown of chlorophyll by the enzyme chlorophyllase which is then followed by the breakdown of the chloroplasts. In chlorosis associated with virus infections, higher levels of chlorophyllase have been detected, the probable enzymatic pathway being:

$$\text{Chlorophyll} \xrightarrow{\text{Chlorophyllase}} \text{Chlorophyllide} + \text{Phytol}$$

Most viral infections that induce chlorosis reduce the efficiency of the residual chlorophyll as well as producing an overall reduction in chlorophyll. With fungal and bacterial infections, the only effect is that of chlorophyll destruction.

The degeneration of chloroplasts in infected plants can be halted by external factors. In tobacco infected with tobacco mosaic virus, for example, large increments of nitrogenous fertilizer can suppress the otherwise drastic loss of photosynthetic tissue.

There are exceptions to the rule that infected plants show reduced photosynthesis. In rust infections of wheat and beans, tissue in the vicinity of the fungal pustules remain green even though the rest of the leaf becomes chlorotic. This **green island** phenomenon can be explained either by assuming that there is selective retention of chlorophyll around the infection sites or, more likely, that the chlorophyll is resynthesized quickly at these sites immediately following the initial degeneration. The mechanism of the maintenance of green islands is unclear although it is thought that the secretion by the pathogen of hormones such as kinetin are involved.

A distinction should also be made between the effects of necrotrophic and biotrophic pathogens. Net assimilation rate is rapidly reduced with necrotrophs as a result of the damage and death of cells. With biotrophs, there is often an increase in photosynthesis immediately following infection, although this depends on levels of infection and cultivar resistance.

Among other effects upon the photosynthetic activities of diseased plants is a disturbance in starch metabolism with, in some instances, a lower starch content in infected compared with healthy leaves after a prolonged period of photosynthesis and, after an equally long period in the dark, a higher starch content in infected leaves. It may well be that starch metabolism, like other photosynthetic mechanisms, varies with the type of disturbance and the type of disease. These mechanisms of metabolic disruption require a much better understanding if we are to fully appreciate their consequential effects upon yield. The effects of pathogens on photosynthesis have been reviewed recently by Habeshaw (1984).

Translocation of Nutrients and Water

The evidence for plant pathogens affecting transport systems in their hosts has been obtained mainly from leaf-inhabiting biotrophic fungi, although similar effects have also be observed with infections caused by necrotrophic pathogens. Of the transport systems in plants, only transport in the vascular tissue and across membranes need be considered. Nutrients and water are transported upwards in the apoplast (xylem) and downwards in the symplast (phloem), and the loading of the phloem is facilitated by transport across membranes.

The overall effect of biotrophic or necrotrophic pathogens is similar even though the former act indirectly by influencing stomatal aperture and root growth whereas the latter produce consequential adverse affects on water and solute movement by killing roots or damaging leaves, often by the production of toxins. With biotrophic fungi such as powdery mildews or rusts, photosynthates originally destined for the development of new shoots and roots, are diverted to sustain the activities of the pathogen, altering the identity of the 'metabolic sinks'. In a similar manner, nutrients such as phosphorus and nitrogen are sequestered by invading virus particles to facilitate replication, this diversion causing deficiencies in the host plant.

The movement of water up the plant in the xylem is due mainly to the 'transpiration pull' caused by evaporation from the leaf surface through the stomata. Vascular wilt fungi can seriously impair this transport, resulting in a net loss of water with the tissues losing turgor and the plant developing the classical wilting symptoms. The mechanism for producing such adverse effects has been closely studied in such fungal pathogens as

Fusarium oxysporum f. sp. *lycopersici*, the causal organism of tomato wilt. The fungus blocks the xylem vessels both by the physical presence of the hyphae and by the production of polysaccharides and pectic products resulting from the activity of pectic enzymes. There is also evidence that the fungus produces fusaric acid and that this disrupts host metabolism by altering the permeability of the cell membrane with a consequent loss of salts and water (Gaümann, 1957). The blockage may be accentuated, paradoxically, by the host's active resistance mechanisms which result in the formation of **tyloses** in the xylem vessels (see Chapter 6).

Ion transport has also been shown to be affected by root-rotting pathogens such as *Gaeumannomyces graminis*, the take-all organism (Clarkson, *et al.*, 1975; Hornby and Fitt, Fitt, 1982). When the fungus is only in the root cortex there is no effect on the movement of water or solutes. When the pathogen invades the vascular system, significant reductions in the uptake of calcium, potassium and phosphate ions have been reported, although Fitt and Hornby (1978) showed that there could also be increased phosphate and calcium in the shoots which, they considered, was probably due to compensatory uptake by other parts of the root system.

Evidence that infection results in reduced phloem export of nutrients has been produced by applying isotopes to diseased leaves where much of them will remain, unlike healthy leaves where they will be transported elsewhere (Owera, Farrar and Whitbread, 1981). Such retention of assimilates has been reported in many biotrophic disease situations with much of the accumulation being within the fungal tissue rather than the host (Thrower, 1965). However, there is very little retention of assimilates with leaves attacked by nectrotrophs. Abnormal phloem transport has also been reported in beans infected with *Uromyces phaseoli* causing nutrients to flow towards the site of infection depleting the surrounding healthy tissue of starch and sugars.

Infection can also cause physical damage to phloem tissue as has been demonstrated in the infection of *Tussilago farfara* by the rust fungus *Puccinia poarum* (Al Khesraji, Losel and Gay, 1980). A number of virus infections are also known to cause a necrosis of phloem elements. A nutrient imbalance results with potatoes infected by leaf roll virus, for example, accumulating carbohydrates in the leaves and having reduced levels in the tubers.

Effects on Growth

The damaging effect of pathogenic attack of plants has been described in Chapter 3 where the many disease symptoms are contrasted. Changes in growth, size or form, can often be directly attributed to loss of photosynthetic tissue due to the activities of the invading organism. Good examples of the relationship of foliar disease and grain yield have been provided by Carver and Griffiths (1981), with a very good correlation of grain yield in barley with the relative green leaf areas when infected with differently timed epidermics of powdery mildew (*Erysiphe graminis hodei*) (Figure 5.1). The effect of vascular wilt fungi on growth has already been referred to in this chapter but, depending on the pathogen and the severity of attack, shortened stems and reduced leaf areas, severe dwarfing, **epinasty** and even plant death occurs. The damage caused by root rot pathogens is just as variable causing many changes in growth. The take-all pathogen of cereals (*Gaeumannomyces graminis*) is capable of completely killing its host or, with later or less severe attacks, produces the classical 'whiteheads' symptoms with empty heads. In Malaya, white root disease caused by *Rigidoporus (Fomes) lignosus* causes more losses in young rubber trees than all other pests and diseases put together (Fox, 1977).

One aspect of pathogenic attack not already discussed is that of the many changes in growth patterns and the production of abnormal structures which are often brought about. Although the mechanisms involved in such changes are not known in every case, there is

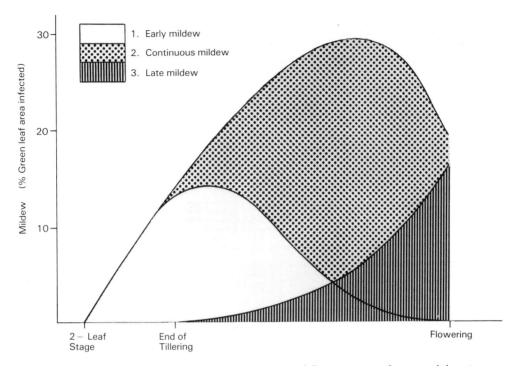

Fig. 5.1 *Powdery mildew* (Erysiphe graminis) *epidemics differing in time of onset and duration.*
A close relationship exists between mildew severity and reduction in green leaf area. Where relative green leaf areas (integrated from emergence to flowering) are plotted against grain yield there is almost complete correlation (r = 0.99). After Carver and Griffiths (1981).

ample evidence to suggest that growth substances (growth hormones) are often involved. There are several detailed reviews on growth substances and plant disease (Sequeira, 1963; Daly and Knoche, 1976; Pegg, 1984), so only a brief consideration of the subject will be given here, with particular emphasis on auxins, gibberellins, cytokinins, ethylene and abscissic acid.

Auxins

Auxin is a generic term for growth-promoting substances based on the indole nucleus and characterized by indol-3-yl-acetic acid (IAA). The latter compound is produced by many pathogenic fungi in culture with diseased plant organs often containing high concentrations. The best evidence to date for a positive and exclusive role for IAA in pathogenesis is work on oleander galls induced by the bacterium *Pseudomonas savastanoi* (Smidt and Kosuge, 1978). This pathogen has also been shown to be capable of converting tryptophan to auxin in pure culture.

The distortions in shepherd's purse (*Capsella bursa-pastoris*) when infected with *Albugo candida* are well-known symptoms of this disease association. IAA levels have also been shown to be significantly higher in infected compared with healthy plants and the distortions can also be produced by the application of the auxin to healthy plants. It should be pointed out, however, that increased IAA levels in diseased tissues do not always cause morphological changes in the host. Higher levels of the auxin are found in wheat leaves infected with *Puccinia graminis tritici* with no associated changes in morphology. In addition, there is considerable variation in auxin content between different tissues of a given host and between hosts of differing disease resistance.

Gibberellins

Although good evidence exists which implicates gibberellins in the host–parasite inter-action, there are many more diseases in which their involvement is less clear. Secretion of gibberellins by some pathogens has been observed since the initial observation by Kurasawa in 1926 of the effect of *Gibberella fujikuroi* in causing the weak and spindly tillers, the bakanae disease, in rice plants. When the active component was isolated, it was named after the fungus producing it and became the forerunner of several other gibberellic acids isolated from diverse micro-organisms and higher plants. There is little information, as yet, on the pathological significance of gibberellins, although striking deformations can be caused by the external application of these compounds. However, exogenously applied gibberellins can reverse the stunting caused by some viruses, suggesting that the virus infection may have caused a decrease in the gibberellin content.

Cytokinins

This is another class of growth-regulating compounds which has been implicated in various aspects of the host–parasite interaction. Cytokinins are known to affect cell division, and enhanced cytokinin activity has been demonstrated in diseased *Vicia* and *Phaseolus* bean leaves infected by the biotrophic rust fungus *Uromyces* spp.

Cytokinin activity is also responsible for the proliferation of lateral buds in sweet peas infected with the bacterium *Corynebacterium fasciens*. Similar increases in cytokynins have been noted in the peach leaf curl disease *Taphrina deformans*, although increased levels of IAA in diseased leaves may also contribute to leaf deformation. The role of cytokinins in delaying senescence is well demonstrated by the green island symptoms associated with the pustules of some rust fungi where the host cells immediately surrounding the sorus eruption remain photosynthetically active. There is also circumstantial evidence that cyto-kinins are involved in the amount of chlorosis associated with some diseases. Certainly, exogenously applied kinetin can suppress the level of chlorosis.

Ethylene

This compound has been shown to be produced by several pathogens in pure culture and has also been detected in a variety of diseased tissues. It has been suggested that the diseased plant produces ethylene as a stress response, its production triggering off many metabolic events which enhance disease resistance. Ethylene is known to activate the synthesis of enzymes such as peroxidase and oxidase and, exposing susceptible sweet potato tissue (*Ipomoea batatas*) to ethylene, certainly increases the activity of these enzymes and induces disease resistance. However, Hislop and Stahmann (1971) could not find a consistent correlation between ethylene production and oxidase and peroxidase activity, and there is always the doubt whether ethylene actually causes these changes in metabolic activity or is the result of these changes.

Abscissic acid

The definitive proof of the involvement of abscissic acid in pathogenicity will be difficult to obtain as it commonly occurs in senescing tissue, affects membrane permeability and acts as an antagonist to cytokinins and gibberellins. Circumstantial evidence comes from the work of Steadman and Sequeira (1970) who found that internode elongation in tobacco plants infected with the bacterium *Pseudomonas solanacearum* was correlated with the production of abscissic acid and the growth of the bacterium, although the inhibitor is not produced in pure bacterial cultures of this pathogen. From this and other evidence it would appear that abscissic acid has a greater significance in plant disease than has been generally recognized, the main effects being stunting, senescence and premature defoliation.

Effect on Yield

When a plant is diseased, many or all of its critical biochemical functions may be affected. The previous sections have discussed effects on photosynthesis, respiration and on growth substances. In general terms, the pathogen normally causes a decrease in CO_2 fixation, an increase in respiration with aberrant starch, amino acid, protein and phenol metabolism occurring. In many host-parasite relationships, these effects are not lethal to the host tissue but, in others, or when infection reaches the advanced stages, one or more of the critical biochemical functions fails and death ensues.

Tissue death may rapidly become organ death with premature loss of leaves in the case of a foliar pathogen. In other host-pathogen combinations, tissue death may lead to destruction of storage tissue as found in tubers, rhizomes and root crops with reductions in not only overall yield but also the quality of the harvested product.

There will, of course, be a threshold level in most crops below which infection will not produce a detectable loss in yield but, it must be emphasized, that many sub-lethal infections or apparently innocuous disease symptoms can so debilitate the host plant that it is incapable of producing its full potential yield.

Reductions in photosynthetic area lower the capacity of the plant to fulfil its functions in terms of yield production and most foliar diseases are in this category. However, pathogens can also have a more direct effect upon yield. Damping-off diseases or diseases which cause the death of seedlings or mature plants will reduce yield proportionately to the number of individuals killed although there may be some compensation from neighbouring plants.

Yield may also be directly affected by pathogens which transform, replace or totally destroy the harvested product. In this category, the ergot (*Claviceps purpurea*), loose smut (*Ustilago nuda*) and choke (*Epichloe typhina*) pathogens are good examples.

Infection can thus be summarized as producing physiological, anatomical and morphological changes any of which, singly or in combination, may result in effects that can cause depression of economic yield.

6

Disease Resistance

Introduction

Bearing in mind the number and variety of plant pathogens, the growing of crop plants would seem to be a most unpredictable economic proposition. However, most plants are resistant to most pathogens and it is the exceptional pathogen:host combination that results in disease of epidemic proportion. A study of any heterogeneous population of plants would reveal that the response to infection is extremely varied with some plants being highly susceptible and possibly being eliminated from the population eventually. At the other extreme, plants may be immune but, in between, there is a continuous range of response. A close inspection of the apparently resistant plants will also be revealing as it would be a rare individual plant that did not exhibit some signs of infection on its foliage, fruit or underground organs. The scientist has been most intrigued by this phenomenon which poses the interesting question of why some infections are limited to give small and localized effects, and others continue to develop and spread often to kill the invaded host.

The mechanisms of resisting most pathogens are in themselves of interest, but to the plant pathologist and plant breeder it is those mechanisms that donate resistance to pathogens that cause economic damage that provide the subject matter of many research programmes.

Following the work of Roland Biffen at Cambridge during the first decade of this century, we know that it is the plant's genotype that determines whether it is resistant or susceptible to infection by a particular pathogen. Over the centuries, plants have evolved in situations where they were exposed to particular pathogens and, by various events which result in the production of new variation and by genetic recombination, certain genotypes have emerged equipped to withstand infection to varying degrees. These events, mutation or hybridization for example, have produced plant characteristics of both a structural and physiological nature which may produce the necessary resistance barrier to a pathogen.

Structural barriers such as the cuticle and epidermal cell wall may provide the first of such obstacles to infection. Physiological resistance may be the result of biochemical reactions to infection or they may be due to the natural chemical nature of the invaded host. In fact, disease resistance in plants may operate at any phase of the infection process, from the time of spore germination on the host's surface, to the colonization of the tissues, to the reproduction of the pathogen on or within its host. Each kind of plant probably employs a different combination of mechanisms depending on the nature of the pathogen and, in general, the hosts' reaction will be modified by the environment. Disease resistance can be studied by investigations of the individual components which contribute to it. However, it is just as rewarding to consider resistance on a crop basis as the presence of large numbers of similarly resistant or susceptible plants can greatly influence the course and severity of an epidemic.

Disease resistance can be studied both from the point of view of its functional operation and from its genetical control. To achieve the best understanding of this subject, a knowledge of the morphology and physiology of the healthy plant is of fundamental importance. The actual mechanisms through which resistance operates must be studied and characterized in detail. The starting point of such a study should, logically, be the arrival of the pathogen spores or other propagules at the host–plant surface. In a susceptible plant, the infection process will continue unhindered but, in a resistant plant, mechanisms limiting the penetration, colonization and reproduction of the pathogen will become functional at varying times in the infection cycle. Some resistant reactions are the result of pathogen-invoked stimulus rather than a pre-existing state in the host plants. The former resistance category has been named **active** and the latter **passive**.

Active Resistance Mechanisms

The first line of defence of plants against pathogens is their surface which the pathogen must penetrate if it is to cause infection. The reaction of a plant to pathogenic attack is often similar to that of the same species to mechanical injury where repair responses might involve the production of substances which act either by sealing off the infected area by initiating cell divisions or by repairing damaged tissues. However, resistance may even commence before penetration is attempted. In many plant species, resistance may be observed at the spore-germination stage where, on resistant plants, germ-tubes may be distorted and grossly inhibited with the overall effect of reducing the success of spores to penetrate the host.

Cork Layers

The formation of a cork layer as an immediate response to penetration has obvious advantages in terms of resistance. Cork layers can be produced as a result of mechanical injury, the cells immediately below the infection initiating a cork cambium layer within a few days of penetration. Cork layers, as exemplified by the reaction of potato tubers to infection by the powdery scab fungus *Spongospora subterranea* (Figure 6.1), are quite common on roots, stems and young fruit and, in addition to preventing further colonization by the pathogen, also prevent damage by the blocking of any toxic substances diffusing from the invading organism. Cork layers are also formed to delimit lesion spread in a number of host:pathogen combinations. A good example is leaf spot of brassicae caused by *Alternaria brassicae* where a cork layer is quickly formed around the infected cells. Eventually, the pathogen breaches this barrier only to encounter another quickly formed cork layer. This process is repeated to produce lesions composed of concentric rings of corky tissue.

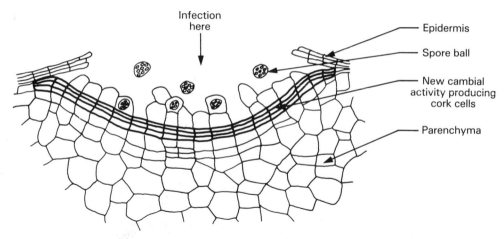

Fig. 6.1 *Diagrammatic representation of the formation of a corky resistance barrier in a potato tuber infected with* Spongospora subterranea

Abscission Layers

A somewhat similar active resistance mechanism is the production of abscission layers in which a cork cambium develops around the infected area and extends from upper to lower epidermis. With the completion of the abscission layer, the centrally infected tissue falls out giving the classical 'shothole' symptoms of plums and cherries infected with *Pseudomonas mors-prunorum* (Plate 6.1). A similar process of lesion restriction occurs in sugar beet leaves infected by the leaf spot pathogen *Cercospora beticola* but here, although the cork cambium is produced to give a restricted leaf spot, the abscission layer does not develop.

Lignitubers and Tyloses

The formation of lignitubers and tyloses are also examples of active resistance mechanisms that either retard the infection process or effect a decrease in the rate of progress of a pathogen through a plant. Lignitubers are produced on the stimulus of the attempted penetration by hyphae of the cell wall. Instead of forcing its way into the cell, the hyphal tip finds that it is increasingly impeded by the deposition of substances, probably cellulose, callose and some lignin, which form a sheath around the invading organism. An example of this type of pre-penetration resistance mechanism has been described by Carver and Carr (1977) with certain cultivars of oats and the powdery mildew pathogen *Erysiphe graminis avenae*. Here, as the spore germinated and produced its infection peg, the host plant formed a **papilla** directly below it which acts as a structural defence barrier (Plate 6.2). Such is also the resistance of certain wheat cultivars to the take-all fungus, *Gaeumannomyces graminis*, susceptible cultivars failing to produce effective resistance barriers. The formation of tyloses in xylem vessels occurs in most plants when invaded by vascular wilt fungal pathogens (Figure 6.2). They are formed by the outgrowth of the protoplast of adjacent parenchymatous cells which then protrude and swell in the vessels often with the effect of completely blocking the spread of the pathogen upwards in the vascular system. The tyloses themselves will also contribute to the wilting symptoms by interfering with the transport of water but they can be very efficient in preventing further advance of the

Plate 6.1 'Shothole' symptoms in cherries (*Pseudomonas mors-prunorum*) (Photo: East Malling Research Station)

pathogen in cultivars in which they are formed quickly enough in the young plant which can also compensate by producing new vessels to replace the blocked ones. Quite obviously, where only partial blocking of the vessels occurs due to slow or sparse tyloses production, such cultivars will succumb to the pathogen. Good examples of tylose forming effective resistance barriers have been described in sweet potato cultivars attacked by *Fusarium oxysporum batatus* and in the hop species attacked by *Verticillium albo-atrum*. Excellent coverage of the role of tyloses and other mechanisms of resistance to vascular wilts can be found in the recent review by Pegg (1985).

Chemical Production

Some plants resist disease by producing chemicals which either form an impenetrable barrier to invading organisms or act in an antibiotic manner. Gum secretion is very common in stone fruit trees infected with bacteria (such as *Pseudomonas syringae*), or in apple and plum trees infected with the silver leaf fungus, *Stereum purpureum*. The gums may be deposited around lesions, in parenchyma and xylem cells as well as intercellular spaces, the susceptible plum cultivar Victoria producing only a small quantity compared with the resistant Pershore in which 'gummosis' is abundant and whose trunks rarely become diseased. The production of resins, tannins and phenolic substances have also been invoked as active resistance mechanisms.

Phytoalexins

Substances acting in an antibiotic manner and only produced in the presence of a plant pathogen have been named phytoalexins. In 1940, Muller and Borger first reported their

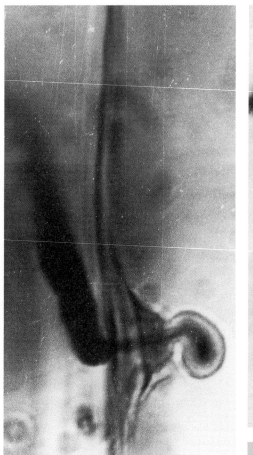

(i)

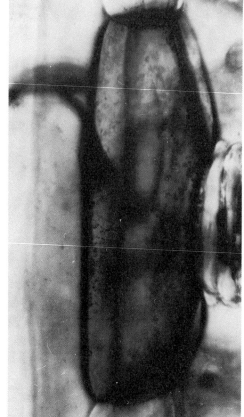

(iii)

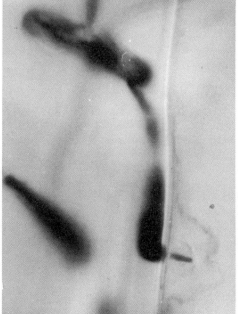

(ii)

Plate 6.2 Plant disease resistance mechanisms
  (i) Encapsulation of infection peg
  (ii) Infection peg halted by papilla formation
  (iii) Hypersensitive cell death after penetration
    (*Erysiphe graminis* on oats)
(Photos: T. L. Carver, WPBS)

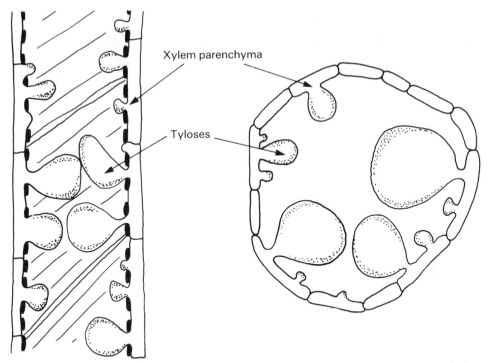

Fig. 6.2 *Vertical and longitudinal sections through xylem vessels showing the formation of tyloses as outgrowths of the xylem parenchymal cells*

presence in potato cultivars inoculated with races of *Phytophthora infestans*. Their name was derived from the Greek and a simple translation would be 'warding off agents in plants'. Their presence has now been established in many crop species although their importance is still something of a controversial topic. Although phytoalexins can be shown to be present, their concentration is often considered too low to account for the observed resistance. However, there is a gradient of phytoalexins with the highest concentrations immediately surrounding the site of primary infection and it is here that the maximum effect would occur.

Chemical analysis of phytoalexins shows that they are organic compounds of varied composition. For example, the compound **phaseollin** was characterized in beans (*Phaseolus vulgaris*) when droplets containing *Monilinia fructicola* were applied. Its chemical formula has been reported by Cruickshank and Perrin (1963):

Phaseollin

It was thought that these compounds might be the products of the fungus itself or as a result of the joint activity of the fungus and the host. However, the fact that phytoalexins could be produced by applying solutions of mercuric or copper chloride showed conclusively that phytoalexins are host-produced.

Another early misconception was that a single host only produced one phytoalexin. The work of Pueppke and Van Etten (1976) clearly indicates, however, that several compounds of differing activity can be produced. In pea pods attacked by *Fusarium solani*, for example, the phytoalexin **pisatin** and three other pterocarpans of similar activity were also elicited and, in potato tubers, **chlorogenic acid, phytotuberin** and **rishitin** have also been identified (Kuć, 1972).

Phytoalexin production in cereals appears to be very limited although they have been reported in maize infected with *Helminthosporium turcicum* and in certain rice cultivars infected with *Pyricularia oryzae*. Also, Mayama *et al.* (1982) identified certain **avenalumin** compounds as phytoalexins in oats infected with *Puccinia coronata avenae*. Interestingly, they also showed that these compounds accumulate in incompatible interactions governed by single major (*Pc*) genes in the host.

The overall picture would seem to be that phytoalexins are formed in many plant species in response to attack by certain pathogens or pathogen races and by chemical or physical wounding. Where the pathogen successfully colonizes the host, it strongly suggests that it can overcome inhibition by any phytoalexin produced. It might well be that, in the susceptible host:pathogen relationship, much smaller amounts of phytoalexin are produced than when incompatible combinations occur. Evidence by Varns and Kuć (1971) confirms this in potatoes in which large quantities of rishitin are produced with incompatible races of *Phytophthora infestans* but not with compatible races.

Other reasons for overcoming the inhibitory effects of phytoalexins are:

(a) differences in sensitivity between compatible and incompatible pathogens, for example, *Ascochyta pisi*, which attacks peas, is relatively insensitive to pisatin whereas it is much more sensitive to phaseollin from French bean, a plant which it does not attack;
(b) the inactivation or degradation of the phytoalexin by the compatible, potentially successful pathogen.

The ability of some parasitic fungi to metabolize phytoalexins has been confirmed by Mansfield and Deverall (1974) who showed that the phytoalexin **wyerone** was degraded by *Botrytis fabae* in bean leaves (*Vicia fabae*), the replacement product being of much lower antifungal activity than the wyerone itself. A recently published book edited by Bailey and Mansfield (1982) comprehensively reviews the topic of phytoalexins.

Induced Resistance

Chemicals are also involved in the phenomenon of induced resistance. The induction of these chemicals may be brought about by the application of an incompatible pathogen. Hammerschmidt and Kuć (1979) reported the isolation of phytotuberin from *Nicotiana tabacum* inoculated with an avirulent bacterium, and also demonstrated that inoculation of one leaf of cucumber with *Colletotrichum lagenarium* systematically protected the plants against disease caused by subsequent challenge with the fungus or *Cladosporium cucumerinum* (Hammerschmidt and Kuć, 1980). They correlated the induced systemic protection with enhanced peroxidase activity and lignification. Other similar examples of induced resistance have been reported as in the reduction of sporulation of one race of *Puccinia striiformis* on wheat after preinoculation with a race to which the host plant was immune (Johnson and Allen, 1975). The phenomenon is likely to be widespread but the biochemical control of this latter example and many others is still being sought.

Hypersensitivity

This is yet another active resistance mechanism which is very common in certain cereal cultivar/pathogen combinations. The hypersensitive response is stimulated immediately after penetration of the host, the first microscopic signs being a discoloration of a few cells in near proximity to the infection site (Plate 6.2). These cells become granular in appearance and then die. Cell death is a successful barrier to further colonization by the pathogen especially if it is a biotroph, requiring living cells from which to extract its nutrient supply.

Externally, the manifestation of hypersensitivity is a flecking of the epidermal tissues. It is a resistance mechanism normally associated with race-specific resistance and is often attributed to single major genes in the host. However, it should be mentioned that some pathologists incline to the view that the cell necrosis is the effect of an undetected resistance mechanism rather than its cause. This view is argued on the basis that, in certain cereal/rust interactions, the pathogen growth rate slows down prior to cell penetration.

Passive Resistance Mechanisms

As with the induced or active resistance mechanisms, passive resistance to disease can be both structural or chemical. The first mechanical barrier to infection is normally the cuticle quickly followed by the epidermal cell wall. Theoretically, a highly waxy cuticle would not only limit retention of water droplets containing pathogen propagules but also impede penetration depending on its thickness. Certainly, leaves of holly and many other highly waxed cuticle species would suggest that this is so. In reality, evidence has only been provided for the *Berberis* spp.:*Puccinia graminis* interaction where there was a clear correlation between the combined thickness of the outer cell wall and a cuticle and the ease of penetration (Melander and Craigie, 1927), and in coffee attacked by the coffee berry disease pathogen (*Colletotrichum coffeanum*) as reported by Nutman and Roberts (1960).

Structural Resistance

The silicon content of rice leaves has also been shown to effect resistance, the silicon being particularly resistant to cell wall degrading enzymes secreted by the pathogen.

Pathogens which invade through stomata may find difficulty in penetration depending on the size of the stomatal aperture. The classical example of this form of resistance is of the resistance to the bacterium causing bacterial canker of citrus fruits (*Pseudomonas citri*). Certain cultivars of mandarin orange with broad lips and small apertures resist invasion; other cultivars and grapefruit are susceptible due to the presence of narrow lips and larger apertures. Similarly, long periods of stomatal closure will reduce infections by *Puccinia graminis* rust spores in certain wheat cultivars although this functional type of resistance may not play as important a role as was suggested at the time of the original observation by Hart (1929). Later work by Romig and Caldwell (1964) showed that resistance in wheat to *Puccinia recondita tritici* was shown to be greater in peduncles and sheaths than in leaves due to stomatal exclusion, resistance decreasing as senescence approached. This resistance varied greatly according to cultivar.

The rate of suberization of lenticels has also been invoked as a preformed mechanical barrier to infection to such pathogens as *Streptomyces scabies* and *Erwinia carotovora* subsp. *atroseptica* on potatoes and *Pezicula malicorticis* (= *Gloeosporium perennans*) on apples. It might be that successful pathogens delay suberization, but Fox, Manners and Myers (1971) have shown that the suberization must be continuous to be effective.

Once the initial structural barriers have been overcome, the pathogen's rate of colonization will depend on the nature of the host's internal tissues. There may be little or no impediment to many pathogens but, in some plant species, certain cultivars have the ability to resist vascular wilt pathogens due to the apparently impenetrable barrier of the endodermis which, in these cultivars, blocks entry to the central vascular system. The student is well advised to read further of resistance mechanisms to these interesting pathogens in the review by Pegg (1985).

Chemical Resistance

Pre-existing chemical substances also contribute to disease resistance. However, the original report in 1929 by Newton, Lehmann and Clarke that phenolic substances accounted for the resistance of some wheat cultivars to *Puccinia graminis* was refuted by Seevers and Daly (1970). There has been no such controversy over the role of catechol and protocatechuic acid in coloured skin onions resistant to onion smudge (*Colletotrichum circinans*). Link, Dickson and Walker (1929) and Wakimoto and Yoshi (1958) found a correlation between phenolic content of rice and resistance to *Pyricularia oryzae*. Preformed factors in living cells of other plants have also been reported. Défago (1977) implicated the saponins in resistance of oats to root pathogens, and avenacins have also been shown to play an important role in oat resistance to *Gaeumannomyces graminis* (Turner, 1961). Chlorogenic acid content is known to affect the resistance of potato tubers to common scab (*Streptomyces scabies*) and, along with phloridizin and arbutin, contributes to the resistance of apples to the scab fungus (*Venturia inaequalis*).

Chemicals are also undoubtedly involved in the specificity of the host:pathogen interaction. It is also generally accepted that 'recognition' of the host by the pathogen, or the reverse, is a major component of specificity. Sequeira (1978) defined recognition as 'An early specific event that triggers a rapid, overt response by the host, either facilitating or impeding further growth of the pathogen'. The whole subject of specificity is very much debated at present and the reader is directed to the book edited by Wood and Graniti (1976). The mechanism of specificity is probably complex and could act at various times in the early stages of germination and penetration. The whole subject has received a major stimulus by the parallel studies of recognition in the legume:*Rhizobium* symbiosis (Dazzo, 1980) in which the lock and key analogy appears to implicate specificity between plant lectins and complementary surface molecules on the bacteria. Attachment of host and invading organism would appear to be an important prerequisite and virulent strains of the tumour-forming *Agrobacterium tumefaciens* have been observed to attach more readily than avirulent strains (Matthysse and Curlitz, 1982). In similar fashion, Kojima and Uritani (1974) have reported the ability of certain plant extracts to preferentially agglutinate spores of non-pathogenic strains of *Ceratocystis fimbriata* whereas pathogenic strains were not agglutinated. It is interesting to note that, in the latter example, agglutination acted as a resistance mechanism whereas, in legumes, specific attachment was between host and compatible strains of *Rhizobium*. This is obviously a topic about which much will be learned in the very near future. In the meantime, the review by Daly (1984) is highly recommended.

Physiological Resistance

If immunity is one extreme of the plant resistance reaction, susceptibility is at the other extreme. The intermediate reactions are, by definition, partial and will vary according to host genotype, pathogen and the environment. **Partial resistance** mechanisms all act at different stages of the disease cycle and all produce effects which, from an epidemiological standpoint, reduce and delay an epidemic. Such mechanisms, affecting incubation period, latent period and sporulation are also termed physiological resistance. **Incubation period** is the time that elapses between penetration and the appearance of visual symptoms whilst **latent period** extends to the onset of sporulation. An increase in either period will reduce the number of disease cycles the pathogen can complete in the season and, coupled with a reduction in sporulation will significantly reduce disease levels. Both incubation and latent period vary with host plant genotype but are also greatly affected by environmental conditions.

Sporulation is very much under the control of the host plant and, often, there are significant host genotype:pathogen isolate interactions (Johnson and Taylor, 1972). Some cultivars of cereals possess a slow rusting character when infected with *Puccinia* spp. the resistance factor being controlled in wheat and barley by genes which produce their effects after the seedling stage.

Other types of resistance which cannot be attributed to any clearly defined mechanisms are **field resistance** and **durable resistance**. Field resistance, often equated with **adult plant resistance**, is considered to be quantitatively inherited, non-specific and probably effected by several mechanisms. Durable resistance, although not always operating at a high level (Johnson and Law, 1973) and with crops occasionally becoming heavily infected, remains effective for the commercial life of the cultivar, 25 years in the case of the wheat cultivar Cappelle Desprez in respect of yellow rust (*Puccinia striiformis*) in the United Kingdom.

Tolerance

The concept of tolerance is based on the assumption that disease and damage are not synonymous. In most crops, the amount of disease at the various stages of its development is quantitatively related to the damage caused in terms of yield loss. With loose smut of barley (*Ustilago nuda*), an infected plant produces no grain so there is no possibility for tolerance but, with diseases which attack foliage to varying degrees, then scope for a tolerance characteristic exists.

If two cultivars appear to be equally damaged by the same pathogen but one cultivar suffers less in terms of yield or quality loss, this cultivar is described as tolerant. Many hypotheses for the mechanisms of tolerance have been advanced but, accepting that it must be a complex phenomenon, only the possibilities that tolerant plants divert plant nutrients from infected to healthy parts, or that certain compensatory factors come into operation, are meaningful explanations. A good account of the tolerance is given by Schafer (1971), who quotes brown rust (*Puccinia recondita*) and glume blotch of wheat (*Septoria nodorum*) as diseases against which some cultivars exhibit tolerance.

Disease Escape

A lesson best learned early in a pathologist/plant breeder's career is that healthy plants are not necessarily immune. Such plants may simply have been missed out in the deposition of pathogen propagules whether under natural conditions or during artificial inoculation. Such rarities occur even with the best of inoculation techniques. In addition, some cultivars of crop plants possess heritable characteristics which enable them to evade infection, possibly by virtue of their development at a time not conducive to pathogen activity or by some morphological feature that inhibits pathogen entry.

The time of maturity of early potato cultivars illustrates one aspect of disease escape, the plants normally being harvested before potato blight (*Phytophthora infestans*) can become established. In the cereals, rye, an outbreeder, is particularly susceptible to the ergot pathogen (*Claviceps purpurea*). Wheat, an inbreeder, opens its florets for a brief period (20–30 minutes) during flowering but the stigma is not receptive to invasion by ergot spores as it will have been fertilized before opening. Barley, also an inbreeder but with a closed flower habit, does not allow entry of ergot spores at any time. This is disease escape although artificial infection of the barley stigma reveals that this host is susceptible and not immune.

Genetics of Resistance

The breeding of disease resistant plants became a science where previously it had been an art with Biffen's (1905) discovery that resistance in wheat to yellow rust (*Puccinia striiformis*) was inherited in Mendelian fashion. By crossing the susceptible Red King cultivar with the resistant Rivet, he produced the classical segregation ratio of three susceptible:one resistant in the F_2 generation. In the F_3, the progeny segregated one homozygous susceptible:one homozygous resistant:two heterozygous segregating lines. He was certainly fortunate in the fact that resistance in this case was governed by a single recessive gene. If this resistance had been polygenically controlled, the genetic basis for resistance may not have been elucidated for many years. In fact, we now know that, in many instances, resistance is dominant and it is now an accepted fact that there is considerable diversity in terms of the genetic control of resistance.

Major Gene Resistance

In many plant genotypes, disease resistance is controlled by only one or at most two or three genes. Such resistance is termed **monogenic** or **oligogenic** and, perhaps because the effect is often dramatic with clearly defined resistance it is often described as major gene resistance. Biffen's early pioneering work with wheat resistance to *Puccinia striiformis* is a good example of this resistance type and the reader can obtain further details of this host:pathogen interaction in the paper by Lupton and Macer (1962), or of the barley:*Erysiphe graminis* interaction in the paper by Moseman (1966).

Monogenic resistance is usually very easy to detect, even at the seedling stage and is mostly specific against one race or a few races of a pathogen. Van der Plank (1963) introduced the term **vertical resistance** to describe this **race-specific** resistance which has been the main objective of plant breeders for several decades, it having the most important attribute of the relative ease of its incorporation into new cultivars. However, experience has consistently shown that, with the exception of the 'slow-moving' soil-borne pathogens, new virulences can quickly appear in the pathogen population that can overcome race-specific resistance with the consequent demise of the cultivar bearing such resistance—the classical *boom and bust cycle* (Figure 6.3). Major gene resistance is often expressed by a hypersensitive reaction and has recently found favour for use in **multilines**, **multigenic** cultivars or **composite mixtures**. This approach is discussed at the end of this chapter.

The specificity of most types of major gene resistance suggests that there is some relationship between the physiological race of the pathogen and the host cultivar incorporating different resistance genes. This specificity would suggest a close relationship between the resistance gene and the gene for virulence against which it acts. This relationship formed the basis of the **gene-for-gene** hypothesis as proposed by Flor (1956). By inoculating numerous flax rust collections (*Melampsora lini*) on to 16 cultivars of flax, Flor was able identify 150 rust races by their differential reaction. By a complex series of crossing experiments coupled with inoculation by different races, he was able to establish that for every gene conditioning resistance in the host, there was a complementary or reciprocal gene conditioning virulence in the pathogen. It is an excellent model with which to study race-specific disease reactions and, although it does not give perfect explanations

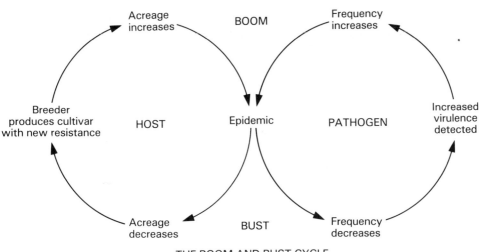

Fig. 6.3   *The boom and bust cycle* (after Priestley, 1978)

for every such host:pathogen system, it has facilitated a better understanding of the genetic basis of resistance.

The gene-for-gene concept also allows for the prediction of the number of races that a set of resistant cultivars incorporating different resistance genes will differentiate (**differential** cultivars). Given that each resistance gene has two phenotypes, resistant or susceptible, if there are n genes for resistance in the host cultivar set, then there would be a possible $2n$ different races. However, this prediction is only valid if each cultivar contains one resistance gene only.

One hypothesis of gene action assumes that incompatibility between the gene for resistance and the gene for virulence is due to the recognition between their respective gene products, incompatibility resulting in a resistant reaction. If no recognition occurs, as when either or both genes are recessive and no gene product(s) are produced which match up, there is no incompatible reaction and compatibility or susceptibility results. This molecular interpretation of the gene-for-gene interaction is depicted in Figure 6.4.

An appropriate example of major gene resistance is that of potato to the late blight pathogen (*Phytophthora infestans*). Major genes for resistance were found at the beginning of the century (Salaman, 1911) in the wild species, *Solanum demissum*. These genes were transferred into the cultivated potato, *Solanum tuberosum* and were designated R_1, R_2, R_3 etc. The race that could attack the cultivar containing R_1 was designated $race_1$, the race overcoming R_2 designated $race_2$ etc. Each of the R genes was shown to be inherited in a simple Mendelian fashion and each gene conferred resistance against the common 'potato race' known as $race_0$. It is now known that in addition to this simple gene-for-gene situation where a race with one virulence is described as a *simple race*, there are cultivars with several resistance genes, the races capable of overcoming them containing several of the corresponding virulences and are described as *complex races*. The Scottish bred cultivar, Pentland Dell for example, contains R_1, R_2 and R_3 but many complex races have been isolated from it, some containing nine virulence genes.

Polygenic Resistance

Where there are several genes controlling resistance, certainly more than three, the resistance is described as **polygenic**. It is generally accepted that, in the main, these genes act in an

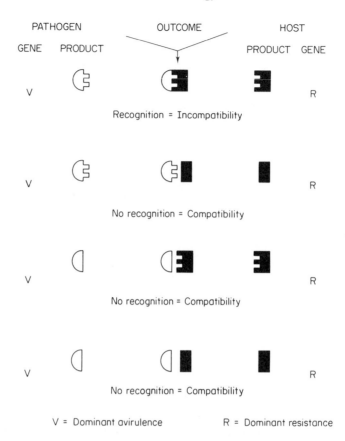

| PATHOGEN | | OUTCOME | HOST | |
| GENE | PRODUCT | | PRODUCT | GENE |

Recognition = Incompatibility

No recognition = Compatibility

No recognition = Compatibility

No recognition = Compatibility

V = Dominant avirulence   R = Dominant resistance

Fig. 6.4 *Molecular interpretation of gene-for-gene interactions between host and pathogen* (after Jones and Clifford, 1983)

additive manner and that they are race non-specific. The level of such resistance will obviously vary with the host genotype and will usually confer a partial resistance which is rarely high. Polygenic resistance is not normally detectable at the seedling stage and is often expressed increasingly in the field as the plants mature.

Polygenic resistance is normally greatly influenced by the environment and it is much more difficult to manipulate in a breeding programme than oligogenic resistance. In addition, complex biometrical techniques have to be employed if genetical analysis is being undertaken.

The number of genes involved suggests that resistance is effected by mechanisms which operate at different times during the disease cycle. These *components of partial resistance* have been discussed already in the section on physiological resistance and their effects can be grouped into three broad categories, those which:

(a) reduce infection frequency,
(b) increase latent period,
(c) reduce sporulation and infectious period.

Much work is at present being done in this important area of resistance and the reader is recommended to read the papers by Parlevliet (1979) and Lancashire and Jones (1985).

Breeding for Resistance

There can be no doubting that the use of resistant cultivars is, potentially, the most cost-effective way of controlling plant diseases. It involves a once only cost to the farmer, which is usually very small, eliminates the requirement for chemicals, and for this reason also keeps the farmer on the right side of the environmental lobby.

We have just seen that there are many and varied ways in which plants can resist plant pathogens and that the mechanisms of resistance are under the genetic control of the host plant. It follows, therefore, that plant breeders should be able to manipulate the host genotype by breeding to obtain the necessary resistance genes in the most desirable agronomic background. The first step in the breeding programme will often be to decide the type and level of resistance required. This will depend upon whether the pathogen is seed-borne, soil-borne, or air-borne, whether it develops early or late and whether it is of regular or sporadic occurrence.

This decision will, of course, depend upon the availability of a suitable source of resistance and whether or not it can be manipulated in a breeding programme.

Sources of Resistance

In the first instance, a search for resistance should be made in the current cultivars of the crop in question. If this is not successful then the search should be widened to screen plant collections, nurseries and the appropriate 'gene-bank'. When such a comprehensive search fails, the breeder has then to investigate related species and possibly progenitors of the crop species.

In this context, there are several public collections available to the breeder. For example, Oxfam recently sponsored a vegetable gene bank at the National Vegetable Research Station, Wellesbourne, Warwick, UK, and there is a very diverse collection of plant species held under the auspices of the USDA at Beltsville, Maryland, USA. The centre for maize germ-plasm and also small grain cereals is CIMMYT, Londres 40, Mexico. These collections are invaluable as sources of resistance to a variety of common diseases and also for resistance to hitherto unimportant diseases. It is important that every attempt should be made to retain and increase these collections for the use of future generations.

In the absence of resistance from conventional sources, it will be necessary to try to create new genotypes by mutation. Much work has been done by the International Atomic Energy Authority (IAEA) in this respect and, although relatively few new cultivars have emerged as a result of 'mutation breeding' the technique is obviously one which should not be ignored. The reader is directed to the publication *The Use of Induced Mutations for Improving Disease Resistance in Crop Plants* (Vienna, 1977). It should be emphasized, however, that breeding for resistance is basically the same as breeding for any other character but, before embarking on a resistance breeding programme, several factors should be taken into account.

(a) the breeding system of the host—inbreeder or outbreeder;
(b) the type of resistance available—active, passive, etc.;
(c) the genetic control of resistance—oligogenic or polygenic;
(d) the genetic variability of the pathogen and its mode of dissemination—physiological races, soil-borne or air-borne, etc.

Characterization of Resistance

The various mechanisms of resistance and their genetic control have already been discussed in this chapter. Information on the mechanism, its site of operation, its specificity and its efficiency is very necessary before critical screening tests of recombinant progeny can be devised. Can the resistance be evaluated in the seedling stage? This is a major logistical problem as the necessity for field testing at an early stage is very expensive. It might be necessary to make two assessments during the course of plant development to both indicate differences in infection rate and to establish any differential resistance with increasing maturity.

After initial screening to identify potential parents for the breeding programme, further detailed characterization to assess environmental effects and further elucidate inheritance will be undertaken as essential preliminaries. Recessive genes are difficult to manipulate in backcross programmes and the complexity of polygenic resistance introduces problems of scale. As has already been discussed, resistance can operate at any stage in the infection cycle and detailed studies will be required to ascertain the relative importance of each mechanism and its genetic control.

Resistance screening must, of necessity, be comprehensive in the first instance as it is easy to discard types that do not express their resistance potential to the full in a highly specific screening test. It may be that the resistance is of the **adult plant** resistance category, in which case, it may be missed or underestimated in a seedling test. Carver and Carr (1977) describe such resistance in oats to *Erysiphe graminis avenae*, the resistance also appearing to be **race non-specific**. Comprehensive screening would also indicate that other adult plant resistances may be **race-specific**, as in some mature wheat plants attacked by *Puccinia striiformis*.

Resistance may be found to be incomplete as in the partial resistance of some barley genotypes to *Puccinia hordei* (Clifford, 1972), or most wheat cultivars to *Septoria nodorum* (Lancashire and Jones, 1985). The nature of partial resistance and its effect in slowing down epidemic rates has already been discussed; suffice to say that any mechanism which reduces spore germination, penetration, establishment, colonization or sporulation should be included in the partial resistance category. The slow development of fewer, smaller pustules in rusts and mildews by **slow rusting** or **slow mildewing** types provide good examples of partial resistance effects on epidemic development.

The variety of host–pathogen resistance responses is considerable. The implication being that it will be necessary to characterize resistance as precisely as possible if the mechanism is to be incorporated in the most effective way. The design of screening methods will be crucial if this is to be achieved with the maximum efficiency in terms of time and labour. Many race-specific resistances will be identified using seedling tests; many of the partial resistance components cannot be accurately evaluated without growing the plants to maturity in the field, and even then comparisons can only be achieved if attempts are made to standardize as many of the technique operations as possible. Environmentally, there will be considerable variation from field to field or from season to season, but some comparisons can be made if inoculation time and inoculum amounts are standard throughout. The numbers of plants normally included in a breeder's segregating population usually mitigate against absolute precision in selection, especially where partial resistance is involved. It is always necessary to subject selected commercial cultivars to a precise final test to ensure that the resistance originally identified in the parental source material is being fully expressed.

Breeding Methods

It is not the intention in this book to delve into the intricacies of breeding methods. The genetical principles and methods used will eventually be the same as in any breeding programme. In practice, the breeder will be endeavouring to incorporate many advantageous characteristics into his new cultivar and disease resistance will be treated no differently from the rest. Many good texts are available on the subject of plant breeding in general, and the reader is recommended to the books of Allard (1960), Williams (1960) and Russell (1978). Much expertise has been accumulated in the breeding of resistance in cereals, and the book by Jones and Clifford (1983) gives comprehensive coverage of this subject.

Traditionally, the **mass selection** or **pure line selection** methods were adopted for heterogeneous populations of plants as, for example, the old land-race varieties. In many respects, it only represented a more rapid method than allowing natural selection to take place and eliminates the more susceptible genotypes.

The **bulk hydrid method** is practised following the hybridization of two parents with useful but different characteristics. After selection, the seed is bulked, grown out again and the process repeated. At each generation, inoculation with a pathogen, either naturally or after artificially applying the appropriate inoculum, allows reselection for the resistant types.

A modification of selection after planned hybridization is the pedigree method which, in cereals, is carried out on individual plants in the F_2 generation when segregation has occurred. The seed from a single head selected in this way is sown out the following year in drills representing the F_3 generation, and the process repeated until F_7 or F_8 when a high degree of homozygosity will have been achieved. The schematic representation of the pedigree breeding method is shown in Figure 6.5.

When the breeder wishes to rapidly introduce a single simply inherited, dominant character into an existing cultivar, he would adopt the **backcross method**, which involves a succession of crossing of the 'donor' plants containing the desirable dominant gene with the existing cultivar. The F_1 and subsequent generation progeny can be tested for resistance and the homozygous recessive and therefore susceptible progeny eliminated. Repeated backcrossing of the resistant progeny on to the recurrent parent (existing cultivar) eventually consolidates the resistance gene in the new genotype. Where the resistance is recessively inherited, it will be necessary to self-fertilize the F_1 progeny and, after progeny testing, select the homozygous plants to use in a backcrossing programme.

Exciting new methods of plant breeding for disease resistance are now being investigated. They can be categorized as involving (a) cell and tissue culture, and (b) recombinant DNA technology. These approaches can all be described as comprising higher plant biotechnology and have transferred much of the preliminary work from the field to the laboratory. However, these techniques must not be considered as replacements for conventional plant breeding and, in general, they will incorporate classical breeding methods, especially in the latter stages.

Cell and Tissue Culture

Meristem culture techniques have long been used in the production of virus-free propagating material. The technique relies on the fact that viruses often do not colonize the apical meristem of such plants as potatoes, tulips, carnations and daffodils. By excising the growing tip and growing it on an agar medium incorporating certain essential nutrients, virus-free plants can be obtained.

Cell culture studies also uncovered another source of variability—the natural variability of somatic cell populations, This **somoclonal variation** gives rise to sports which are vegetatively reproduced and, often, such variants will possess improved resistance characteristics. Shepard, Bidnay and Shahin (1980) demonstrated improved disease resistance in potato plants regenerated from tissue culture, and Larkin and Scowcroft (1981) regenerated sugarcane clones resistant to *Helminthosporium sacchari* from single-cell populations of a susceptible clone treated with helminthosporoside. In some breeding programmes, isolated protoplasts are being used in culture rather than whole plant cells. Using such a method, Jellis, Gunn and Boulton (1984) regenerated isolated protoplasts of three potato cultivars and found improved resistances to common scab (*Streptomyces scabies*) potato virus Y and potato leafroll virus.

68

Plant Pathology

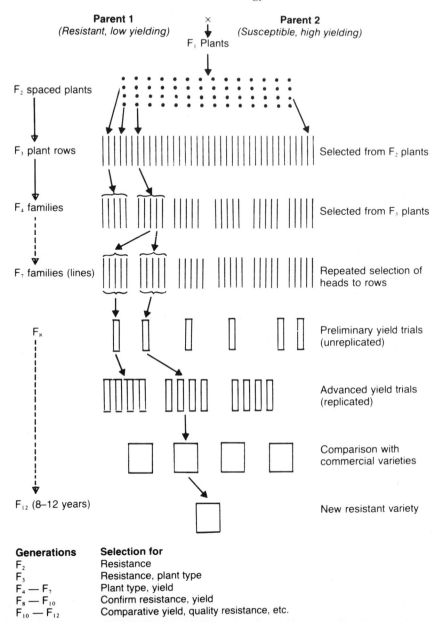

Fig. 6.5 *Schematic representation of the pedigree breeding method* (after Jones and Clifford, (1983))

Mutations have also been induced in tobacco callus culture using a variety of mutagens, the new variants possessing improved resistance against certain pathogens. Possibly the most promising of the new methods has been the so-called 'haploid breeding' technique (Vasil, 1980). This is, in effect, a modification to the pure-line or pedigree method of plant breeding. The initial crosses are made and, following segregation and some selection, pollen grains are selected from the F_2 or F_3 generation. These are, of course, haploids and after a period of culture on a suitable agar nutrient medium, the cells are induced into the diploid state, mostly by treatment with the chromosome doubling chemical colchicine. Plantlets can then be regenerated and incorporated back into a conventional pedigree-type programme. In some plant species, the brassicae for example, the haploid cell culture regularly reverts to the diploid state by spontaneous doubling-up of the chromosomes without the addition of colchicine.

The double haploid technique has also been used to produce commercial cultivars of barley. In this species, use is made of *Hordeum bulbosum* to cross on to segregants in a crossing programme. The embryo of such crosses are initially diploid but quickly lose the *H. bulbosum* chromosomes leaving a haploid set of recombinant chromosomes from the original parents. The young plantlets produced from such an embryo are then doubled with colchicine and the remaining breeding carried out conventionally.

Recombinant DNA Techniques

The purist would state that any breeding method could be classified under this heading. For convenience, the classical methods based on hybridization have already been discussed but some recent techniques involving alien gene transfer can be used as examples of the vanguard experiments leading to more sophisticated recombinant DNA techniques. Alien germplasm can be introduced into a cultivated species at several levels. In Triticale, an entire genome was added by crossing tetraploid or hexaploid *Triticum* spp. with *Secale cereale*. At a second level, alien chromosomes may be added to or substituted for specific chromosomes of the cultivated species. Even more precisely, only part of the chromosome is transferred.

Most of this type of work has involved the transfer of disease resistance to the bread wheats (*Triticum aestivum*), but the reader is referred to a recent review of this subject (Knott and Dvorak, 1976). Alien gene transfer, or more exactly chromosome substitution was effected in *T. aestivum* using a gene situated on chromosome 5B in the wheat genotype which controls homologous pairing. A genotype of *Aegilops speltoides* will suppress the action of this gene and allows genetic exchange between homoeologous chromosomes. In this way, resistance to yellow rust (*Puccinia striiformis*) was transferred from *Aegilops comosa* to *T. aestivum* by Riley, Chapman and Johnson (1968). A schematic representation of the protocol is shown in Table 6.1.

Recombinant DNA techniques today invoke the description 'genetic engineering'. In simple terms, desirable genes may be excised from one genome by the cutting action of restriction enzymes. The genes can be cloned in a suitable vector, predominantly a pro-karyote, and then transferred to the new host. The first successful transfers of genes have recently been made using the bacterium (*Agrobacterium tumefaciens*) as a vector with the genes being carried on the tumour-inducing plasmid (*Ti* plasmid) which is injected by the bacterium into the host cell (Murai *et al.*, 1983). As yet, no disease resistance genes have been transferred, but it is still very early days and much work is in progress to identify, isolate or synthesize resistant genes. An excellent review of unconventional methods of breeding for resistance has recently been published (Wenzel, 1985).

Strategies for the Use of Resistant Cultivars

Although the breeding of resistant cultivars is of relatively recent origin, it soon became apparent that their longevity would not always be as good as might have been first expected. The commercial life of most wheat cultivars is often around four years, by which time more productive cultivars would have superseded them and, in the case of cultivars with race-specific resistance, new virulence genes would have appeared to overcome the resident resistance. Inevitably there will be exceptions and the commercial life of some root crops resistant to soil-borne pathogens are good examples.

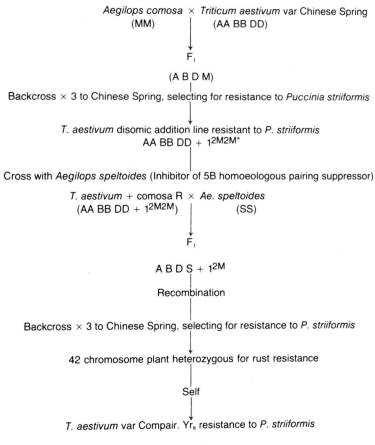

Aegilops comosa × *Triticum aestivum* var Chinese Spring
(MM)   (AA BB DD)

F₁

(A B D M)

Backcross × 3 to Chinese Spring, selecting for resistance to *Puccinia striiformis*

T. aestivum disomic addition line resistant to *P. striiformis*
AA BB DD + 1²M2M*

Cross with *Aegilops speltoides* (Inhibitor of 5B homoeologous pairing suppressor)

T. aestivum + comosa R × *Ae. speltoides*
(AA BB DD + 1²M2M)   (SS)

F₁

A B D S + 1²M

Recombination

Backcross × 3 to Chinese Spring, selecting for resistance to *P. striiformis*

42 chromosome plant heterozygous for rust resistance

Self

T. aestivum var Compair. Yr₈ resistance to *P. striiformis*

* Substitutes for wheat chromosomes 2A, 2B or 2D.

The outlook for the use of race-specific cultivars might appear bleak, but there are situations when they may be utilized to advantage and current methods of resistance diversification are attempts to prolong the useful life of such cultivars by presenting the pathogen with a more difficult barrier to overcome.

Diversification of resistance can be put into operation at a number of levels: regional, local, farm, field or crop. Regional diversification requires much co-operation, possibly between countries, states or other administrative areas. It might require legislation to achieve the maximum efficiency. Van der Plank (1963) has suggested that diversification of resistant genes in bands or regions across the '*Puccinia* pathway' in North America would effectively eliminate any inoculum generated in one band or region when it was deposited on cultivars possessing different resistant genes in the next band downwind in the pathway.

On a country, state or even locality basis, deployment of different resistances in winter-sown and spring-sown cereals might also reduce the epidemic potential. The theory is that virulences against resistances in the spring-sown cultivars would be at a low frequency in the population being transferred from winter-sown cultivars. Similarly, virulences against 'winter-resistances' would decline in the summer population as they would have no selective advantage in that population.

On a farm scale, spatial diversification should be practised. Its success would be difficult to prove in the statistical sense but, theoretically, such diversification should succeed as the majority of air-borne inoculum moves only a short distance from its source. It has been estimated, for example, that 99% of cereal rust urediospores are deposited within 100 m of the source. Any new inoculum emerging in the population would be kept at a low level both by the spatial separation of susceptible plants and by the diluting effect of the various resistance gene barriers.

To facilitate farm diversification, the United Kingdom Cereal Pathogen Virulence Survey initiated a scheme for wheat cultivar deployment. It recommends that at least three cultivars are grown on the farm, chosen in such a way that the risk of disease spreading from one cultivar to another is reduced. The schemes should be used to select suitable combinations of cultivars to sow in adjacent fields or in the same field in successive years. Cultivars may also be selected from the scheme for use in seed mixtures (see below). The symbols in the diversification (Table 6.2) indicate the risks involved in growing any particular pair of cultivars together. Wherever possible, variety combinations marked ' + ' should be chosen, since there is only a low risk that disease will spread from one cultivar to the other.

If a mixture is to be grown, the following guidelines should be used:

(1) Choose component cultivars using the diversification scheme.
(2) Ensure that the component cultivars are not all susceptible to another disease.
(3) Choose high yielding component cultivars.
(4) Avoid extreme differences in earliness of ripening of the component cultivars.

Mixtures and Multilines

The history of mixtures within and between species goes back centuries, their purpose being to help protect crops against stresses. Mixtures are widely used in a variety of crops and a recent review by Wolfe (1985) describes their use on cereals, legumes and potatoes in countries as far apart as Pakistan, Canada, Russia and many other countries in Africa, Asia and South America. Tozzetti is given the credit for first reporting the use of mixtures to counter disease (see Groenewegen and Zadoks, 1979) in his account of the reduction in rust infection in mixtures of wheat and oats.

The success of heterogeneous plant populations in reducing the rate of increase of disease in a crop depends on several factors, not least being the choice of components for the mixture or multiline as has already been discussed. In the first instance, the success of the incoming inoculum to initiate an epidemic will depend upon the proportion and distribution of susceptible plants in the crop. In a homogeneous crop, any plants infected would soon generate inoculum and this would very quickly infect neighbouring plants. In a heterogeneous crop, any spores generated would have a reduced chance of infection (a) in proportion to the number of susceptible plants; (b) due to the diluting effect of the resistant plants in the population which would act as physical barriers to spore movement, the spores being trapped on such plants but not being able to infect them. The physical effects of the mixture or multiline would thus be both isolation and dilution.

It has also been suggested that heterogeneous crops present a physiological barrier in that a **cross-protection** mechanism may operate. Cross-protection can occur when normally susceptible genotypes acquire resistance following prior infection with avirulent pathotypes which trigger a general resistance response.

A **multiline** is a combination of isogenic lines, identical in all agronomic characters but differing in race-specific resistance. Each component line is produced by a programme of recurrent backcrossing of a resistant source on to a common parent. The component lines can then be assembled in varying proportions to produce the multiline cultivar. Multilines have been successfully used to control crown rust (*Puccinia coronata*) in Iowa and to control wheat stem rust (*P. graminis*) in Colombia. Their history and use has been described in a recent review (Browning and Frey, 1981).

Plant Pathology

Table 6.2 Cultivar diversification scheme to reduce the spread of yellow rust (*P. striiformis*) and powdery mildew (*E. graminis*) of wheat.

Companion Diversification Group

	1	2	3	4	5	6	7	8	9	10
1	+	+	+	+	+	+	+	+	+	+
2	+	m	+	+	+	+	+	+	+	+
3	+	+	m	+	+	+	+	+	+	m
4	+	+	+	m	m	+	+	+	+	+
5	+	+	+	m	m	+	+	m	+	+
6	+	m	+	+	+	ym	m	+	+	+
7	+	m	+	+	+	m	ym	+	+	+
8	+	+	+	+	m	+	+	ym	+	+
9	+	+	+	+	+	+	+	+	y	y
10	+	+	m	+	+	+	+	+	y	ym

Chosen Diversification Group

+ = low risk of spread of yellow rust or mildew

y = risk of spread of yellow rust

m = risk of spread of mildew

Each Diversification Group comprises cultivars possessing different resistance genes to the two diseases. If more than one cultivar is to be grown on the farm, the companion cultivars should be chosen from different Groups. It is recommended that low risk combinations should be grown in adjacent fields or as cultivar mixtures.

(Based on the winter wheat Diversification Scheme as proposed by the United Kingdom Cereal Pathogen Virulence Survey and reproduced in modified form with kind permission)

A cultivar **mixture**, or blend, is simply compounded by mixing cultivars on the basis of their predicted performance using the diversification schemes. It is important that each component cultivar is as agronomically similar as possible.

Diversification of resistance obviously presents the pathogen with a much more difficult target than in the traditional monoculture of homogeneous crops. There is always the fear however that with so many resistance genes confronting the pathogen population, a 'super race' might evolve which would have the necessary virulences to overcome all the resistance present. The emergence of 'super races' has been reported but their build-up has not occurred to any extent. This is probably due to a lack of 'fitness' associated with the possession of multiple virulence genes. This may not always be the case although most pathologists are inclined to the view that given enough host diversity, the pathogen's self-regulatory mechanisms operate so that 'super races' need not be feared.

7

Pathological Techniques

Introduction

There is much detail to be studied if the aspiring pathologist is to become an accomplished diagnostician in the field. We have already considered symptomatology in Chapter 3 but, to obtain a comprehensive knowledge of disease development within a crop, it is also necessary to study the causal organism separately. Ideally, this will entail its isolation into pure culture, a complete knowledge of its morphology and life-cycle as well as aspects of its physiology and effects of, and possibly interactions with, various nutrients and environmental factors which may affect disease epidemiology. In brief, the complete pathologist must become adept at handling plant pathogens in the laboratory and this will involve him in many and varied techniques which are discussed below.

The laboratory furnishes the pathologist with the opportunity to make a more precise study of diseased plants, to record his observations with drawings and descriptions, to study the pathogen in its intimate relationship with the infected host plant and to make a more detailed study of the pathogen microscopically and in pure culture. Eventually, the pathologist has to master techniques or storage and inoculum production if he is to progress with the protocol of Koch's postulates. The study of diseased plants under laboratory conditions may often be criticized for its artificiality but, conversely, it can often provide the stable and uniform conditions to solve certain problems of the infection process, elucidate resistance mechanisms and develop and evaluate inoculation methods.

Isolation

It is a prerequisite for the precise study of most plant pathogens that the causal organism is isolated and grown in pure culture. This is no problem with most fungi and bacteria but, even approaching the twenty-first century, there is still no satisfactory way of culturing the obligate parasites such as the rusts, downy and powdery mildews.

It is not always necessary to isolate a fungal pathogen if the aim is only one of identification. Often, many fungi can be identified by their fruiting bodies, pycnidium or ascocarp for example, which may be produced on the diseased plant but not necessarily so in pure culture or artificial media. With those fungi that can be cultured, the first step is their isolation from the diseased host material. Unfortunately, plants are very rarely invaded by a single micro-organism and contamination by saprophytic fungi and bacteria poses the most difficult initial problem. Careful selection of the host material will help to minimize this problem and there are several surface disinfecting techniques which will also help.

Direct Isolation

In its simplest form, direct isolation involves the transfer of visible propagules of the pathogens, such as spores or fruiting bodies, to an artificial growth medium. This will normally be an agar medium in a petri dish containing the necessary nutrients for the growth of the particular pathogen. In some instances, the medium may be highly specific for the organism to be isolated, the **selective** medium. More usually the medium will be of a general nature, able to support a wide variety of fungal or bacterial species. The medium may be acidified with lactic acid to reduce bacterial growth, or rose bengal added with the same purpose. Where spores can be picked off directly, it is normal to suspend them in a drop of sterile water which is then streaked out over the agar surface to produce separate colonies.

It is sometimes necessary to pretreat the diseased tissue to stimulate sporulation prior to isolation. This may be done by placing the host material in a moist chamber which may be simply a plastic bag containing moistened cotton wool or, slightly more sophisticated, a plastic container lined with moistened blotting paper. Special attention should be given to ensure that the moisture chambers are kept at the optimum temperatures for fungal growth and sporulation. It might also be necessary to provide light of a specific wavelength, NUV (near ultraviolet) often being used.

For fungi which sporulate poorly, a good technique is to induce mycelial growth by surface sterilizing the infected tissue by dipping in 1% sodium hypochlorite and then transferring a small piece of tissue, which is dissected out with a flamed scalpel, to acidified agar; potato dextrose agar (PDA) is commonly used. When the mycelium grows out into the agar, hyphal tips from the leading edge of the colony can then be transferred to a fresh agar plate, normally PDA with no added acid.

Dilution Methods

With bacteria and fungi which sporulate profusely, dilution methods may be used. The **plate dilution technique** simply involves diluting the infected material into a medium. By this means, a quantitative and qualitative assessment of the presence of the organism may be made. The method is particularly appropriate for soil-borne pathogens where a soil suspension can be prepared and dilutions made 1:10, 1:100 etc. Aliquots (1 ml) of each dilution can then be incorporated into a molten but cool agar (45 °C) and allowed to set and then incubated. With bacteria, loopfuls of the various dilutions can be streaked on to agar plates. The plates should be labelled before streaking with the suspensions and the agar dried with their lids slightly raised in an oven or incubator running at 25–30 °C.

Soil Plate Method

A modification of the plate dilution method is Warcup's soil plate method (Warcup, 1950). This involves crushing a small amount of soil (5–15 mg) in sterile water in a petri dish. To this is added 10 ml of cool, molten agar, usually acidified to pH 4.0 with phosphoric acid to inhibit bacterial growth. This method has the advantage that organisms which occur within soil aggregates will appear on the plates. Many of these might well not appear in soil dilution plates. Where soils are expected to yield high numbers of micro-organisms, the number of colonies appearing per plate can be reduced by diluting the soil with sterile sand prior to crushing.

Baiting Methods

The use of baiting methods is common when specific fungi are being isolated. These methods vary from the use of appropriate nutrients to the use of plant tissue or even whole plants as 'bait'. With *Gaeumannomyces graminis* for example, a susceptible cereal can be sown in contaminated soil and the fungus eventually isolated from the infected roots. For the isolation of *Thieleviopsis basicola*, slices of carrots are placed on the soil sample in petri dishes and the fungus isolated from the carrot 'bait'. With zoosporic species, various bait, consisting of pieces of appropriate plant tissue, are placed in the soil and, at intervals, samples are withdrawn and transferred to a container containing sterile pond water and distilled water (1:2) and fresh bait from which the pathogen can be isolated by transfer to an agar medium.

Surface Sterilization

For the isolation of bacterial pathogens, the manner of collection of diseased material is crucial. Plastic bags should be avoided as these promote the growth of secondary sapro-phytic organisms. Dried plants samples are the best and these can be obtained by placing the plant tissue between layers of absorbent paper until required. It will usually increase the efficiency of isolation if the host tissue is surface sterilized. This can be achieved with a 2 minute immersion in for example, a 1% sodium hypochlorite solution followed by several washings in sterile water. However, it should be emphasized that with some leaf tissue there is a danger of the chemical being absorbed and killing the bacteria within. In such cases, it is better simply to wash the leaf surface in running tap water.

After sterilization, small pieces of infected tissue, such as the margin of leaf spots including some healthy tissue, are placed on a sterile microscope slide in a drop of sterile water. A flamed scalpel is then used to cut into the tissue and, after about 1–2 minutes, a loopful of the water can be streaked out on to nutrient agar, the bacteria having diffused out of the tissue into the suspending water. With woody species or seed-borne diseases, it is usual to macerate the tissue after surface sterilization which, in these cases, can be achieved by dipping in alcohol which is then ignited to burn off the sterilant. Maceration may only be a tearing of the infected tissue on a glass slide with sterile needles or scalpels.

Single Spore Isolation

For genetical studies or when the investigation demands absolute purity of a culture, it will be necessary to carry out single spore isolation. A simple approach, particularly relevant to bacteria, is to suspend a cell suspension into molten (45 °C) agar before pouring the plates. Single colonies are then picked off and a new suspension made and the whole process repeated five times when there is a high degree of certainty that the resulting colonies will be **clones**.

With fungal spores, a suspension or dry spore cloud from infected plant tissue is allowed to settle onto a stiff (3%) water agar. Given sufficient time to germinate, single spores can then be located using a microscope and either picked off manually using a flattened platinum needle or by lowering a special microscope objective on to the agar surface the lower edges of which cut a circle around the individual spore. Once marked, the small agar discs can be transferred to fresh agar plates containing the necessary nutrients

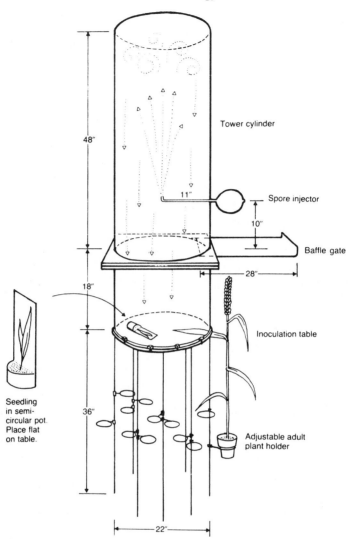

48″

11″

10″

28″

18″

36″

22″

Tower cylinder

Spore injector

Baffle gate

Inoculation table

Seedling
in semi-
circular pot.
Place flat
on table.

Adjustable adult
plant holder

Fig. 7.1 *Schematic representation of a spore settling tower* (after Eyal, Clifford and Caldwell, 1968; redrawn in Jones and Clifford, 1983).

for the growth of the particular fungus. Single spores can also be moved around on an agar plate using a 'micro-manipulator' operated by a 'joystick' mechanism whilst monitoring the spore's movement through a binocular microscope.

With obligate parasites, direct inoculation of plants or plant tissue is necessary. With *Erysiphe graminis*, a single conidial chain can be used, each spore in the chain being assumed to be identical. Alternatively, spore:talc mixtures may be used to inoculate plants using a settling tower (Figure 7.1) to obtain well-spaced infections.

Leath and Stewart (1966) describe a method in which single pustules of rust pathogens (*Puccinia* spp.) on fragments of plant tissue are washed in the surfactant Tween 20, rinsed in distilled water and transferred to agar containing benzimidazole (60 ppm) in the dark for about 4 hours followed by transfer to the light for 2 hours. Any spores present will germinate and desiccate and they can be collected using a Tervet cyclone collector after about 48 hours (Tervet *et al.*, 1951).

Culture and Inoculum Production

In general, the culturing of plant pathogens is carried out on solid agar medium. Often an 'all-purpose' type of medium can be used on which a great variety of fungi or bacteria can be successfully grown. Potato dextrose, corn meal, oat meal and malt extract are commonly used for fungi whereas nutrient or beef peptone agar are good media for bacteria. Broth cultures are normally only used for the culture of bacteria but almost all phytopathogenic bacteria will grow quite adequately on solid nutrient agar medium. Broth cultures, however, have the advantage of facilitating easy preparation of series dilutions of cell suspensions either for precise inoculation studies or other experimental purposes.

Obligate parasites, such as the rusts and powdery mildews, pose a problem in that they require living host material for their sustenance. Some success has been achieved by the use of defined media in certain rusts (Raymundo and Young, 1974) but, in general, living host plants must be used or, over a short period, detached leaf segments may be used if floated on a solution of the senescence retarder benzimidazole (20–100 ppm). Detached leaves may also be placed on weak agar (0–5%) with added benzimidazole, their cut ends inserted into the soft surface, or they may be placed on blotting paper soaked in the senescence retarder, their extremities held down by strips of water agar. This technique is used extensively for the maintenance of rusts and mildews, green leaf being maintained for up to 14 days or so, the fungi lasting for several weeks. Regular transfer of the pathogens to fresh leaf segments is necessary and makes the study of obligate parasites very labour-intensive.

Fungal cultures can be maintained on solid agar for many months if stored at 2–5 °C. Drying out of the agar will eventually occur and a constant watch should be kept so that subculturing may be carried out before the desiccation and possible death of the pathogen. Long-term storage and repeated subculture is often accompanied by a loss in pathogenicity or, at least, a reduction in virulence. Except where pathogenicity is completely lost, a single passage through a susceptible host plant will often revive the culture to its original vigour.

With such important disadvantages, it is hardly surprising that modifications to this simple and cheap maintenance technique have been sought. The most common of these are covering the agar cultures with oil, often glycerol or refined paraffin oil. Fungal cultures in broth with 15% glycerol and bacterial cultures in phosphate buffer plus 15% glycerol can be successfully stored for long periods at −20 °C. Other methods involve storing the pathogen in sterile soil or sand and by freeze-drying or **lyophilization**. The latter is probably the most widely used method for storing bacteria. The method relies on the fact that a bacterial suspension will suddenly freeze if subjected to a high vacuum (around 10^{-3} mmHg) and after this will continuously dry in 5–6 hours. The freeze-dried cultures can be stored in small, sealed glass ampoules and can retain their viability for at least 15–20 years. Dry spores of many rust fungi can be stored without treatment in a domestic refrigerator for up to one year.

The production of spores by fungi on artificial media often depends upon both the amount and quality of light. With both *Septoria* spp. and *Pyrenophora* spp., sporulation may be enhanced by incubation under near ultraviolet (black or NUV) light, although the vast majority of fungi will produce adequate spores in the dark (Plate 7.1).

Large quantities of *Puccinia recondita tritici* uredospores can be collected on V-shaped paper strips placed between rows of wheat seedlings grown under a controlled environment in a 'rust factory' (van der Wal and Zadoks, 1976), or they can be collected from infected plants using a small cyclone collector which acts like a miniature vacuum cleaner (Tervet *et al.*, 1951).

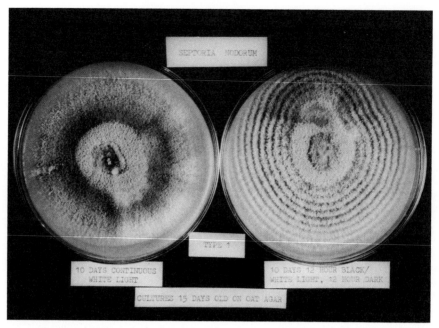

Plate 7.1   The effect of near ultra-violet ('black light') on sporulations in *Septoria nodorum* (*Leptosphaeria nodorum*). Note concentric circles of vegetative myceliums alternating with dark pycnidia corresponding to the white/black light periods (Photo: E. Wintle)

Inoculation Methods

Most plant pathology studies, whether in the laboratory or the field, eventually require the inoculation of healthy host plants. The aims of such experiments may be the establishment of the proof of pathogenicity by Koch's postulates, the study of disease **aetiology** or perhaps the screening of fungicidal chemicals. Artificial inoculation of plants for these and other purposes can be achieved using a number of techniques.

The direct application of plant pathogens can be undertaken by spraying on to the host plants either fungal or bacterial spore suspensions. There are many sprayers which can be used for this purpose, ranging from a simple scent spray to a hand-pump operated portable garden sprayer. Precise application can be achieved by using micropipettes. It is a normal practice to add a few drops of a surfactant, possibly Tween 20, to the suspension to facilitate better contact with the host surface. Gelatin or agar may also be added to help the spores stick to the host surface (Plate 7.2). This method has the advantage of combining the deposition of the pathogen propagules with the provision of adequate moisture for their germination. Fungal spores may also be applied dry when mixed with talc, the use of a spore settling tower allowing even distribution of spores. Dry spores can also be applied using an artist's paint brush, simple ladies powder puffs or insufflators.

A simple alternative method is to shake infected sporing plants over the healthy plants or infected debris may be spread amongst the healthy growing plants. An interesting technique was developed to inoculate wheat or barley seedlings with the eyespot fungus (*Pseudocercosporella herpotrichoides*). Here, straws are inoculated with the pathogen and, after a period to allow colonization, the infected straws are placed over the emerging coleoptiles in the field, the pathogen easily passing from the straw to the seedling. The use

Plate 7.2 Artificial inoculation of barley plants using an atomized spore suspension (Photo: E. Wintle)

of a highly susceptible cultivar which is inoculated at an early stage can provide a potent inoculum source if planted at intervals as **spreader rows** within the crop to be infected.

Wounding

Weak parasites may require a more drastic pretreatment to facilitate entry into the host. A wounding technique may be used to simulate entry points normally produced by insects or other animals. Such wounds can be produced by puncture using a sterile needle or the host surface may be gently rubbed with a mild abrasive such as carborundum. Wounding techniques are almost always required to ensure successful entry of bacterial pathogens. Carlson *et al.* (1979) has described a hand-held pressure injection device which is commonly used for the inoculation of maize with *Xanthomonas stewartii*.

Soil Infestation

With root-infecting fungi, soil infestation methods can be employed. These can range from the mixing of a naturally infested soil into the plant culture medium (soil) or the addition of infected plant tissue into the rooting medium. The take-all fungus, *Gaeumannomyces graminis* can be grown up on a sand and corn meal medium and then added as a 10% constituent of a white silica sand medium.

Use of Vectors

The role of vectors in the infection process of bacteria and viruses has already been discussed (Chapter 2). Viruses, in general, have evolved specific relationships with particular insect, fungal or nematode vectors; *Erwinia stewartii* for example is transmitted by the corn flea beetle, *Chaetocnema pulicaria*. Bacteria, on the other hand, have no such specific vector relationships, the vectors carrying the pathogen cells passively as a result of chance contact.

Disease Assessment

The assessment of the amount of disease on a plant, more importantly on a crop, is essential in any quantitative epidemiological study. Disease assessment also forms the fundamental basis of many other aspects of plant pathology. Assessment data are essential to breeders, fungicide manufacturers, economists, government agencies and academics in their various evaluations of resistance, treatment efficacy and resource priorities.

With such diverse investigational objectives, the precision of the assessment methods and the quantity of data collected will vary accordingly. Inevitably, as large areas may be involved, sampling will be carried out, a compromise being necessary between what ideally should be done and what it is practical and economical to do.

Most importantly, assessment of a particular disease in an individual crop over several years can provide indicators to elucidate the factors governing its incidence and severity. This information can also be used to devise forecasting systems as is discussed in Chapter 8.

Assessment Methods

Over the past thirty years or so, especially since the introduction of computerized analyses, the methods of recording disease have changed from the previous subjective, somewhat descriptive 'slight', 'moderate' or 'severe' to a more standardized, often pictorial type of assessment key which is also quantitative. Assessment methods may be direct or indirect, quantitative or qualitative. In its simplest form, only the absence or presence of disease may be recorded as **incidence.** More usually, it is the **severity** of disease that requires measurement but, especially where responses to individual virulences (physiological races) are under scrutiny as in breeding programmes or race surveys, a qualitative method of assessment may be used. An example, as used in the differentiation of cereal rust virulence responses, is shown in Table 7.1 below. The convention is to group 0-, 1- or 2-type responses as resistant and 3- and 4-types as susceptible.

The nature of the disease and its distribution in the crop will determine the assessment method. In general, each disease demands its own assessment key but, as in the case of cereal rusts, a standard key may cover several rust species. There are basic requirements for each and every method. They must be **rapid, reliable** and **reproducible** not only from one worker to the next but from one location to another and from season to season.

Table 7.1 *An Assessment Key for Cereal Rust Virulence Response*

Symbol	Host: parasite interaction
Oi	Immune; no visible signs of infection
Oc	Highly resistant; minute chlorotic flecks
On	Highly resistant, minute necrotic flecks
1	Resistant; small pustules with necrotic surrounding tissue
2	Moderately resistant; medium-sized pustules with necrotic surrounding tissue
3	Moderately susceptible; medium-sized pustules with chlorotic surrounding tissue.
4	Susceptible; large pustules with little or no chlorosis
X	Mesothetic reaction; mixed reaction types on one leaf

Outbreaks of disease in relatively large perennial plants, such as apples, bananas and forest trees, may be monitored visually at regular intervals. With annual crops, where plant death can occur rapidly, there is an even greater requirement for regular inspection in order that the danger threshold can be identified and a control treatment applied. Where a disease increases in extent during a single season, it will probably be necessary to make several assessments, geared to physiologically significant plant growth stages. The rate of increase of disease may also be calculated by regular interval assessments.

Disease assessment keys have now been constructed for many pathogens and their regular use and citation in the literature is evidence of their usefulness and general acceptance. With many keys, the operator will be able to compare a particular sample of diseased plants with an accurately drawn pictorial depiction of the normal range of disease responses. He will be able to scan many plants quickly by eye and, with the aid of a portable tape recorder, capture the data rapidly in a form suitable for transcription to a more permanent form on his return to the laboratory.

There will be difficulties with the assessment of blotches, stripes and mosaics but the standard keys, a selection of which is shown in Figure 7.2, can produce very accurate assessments if the operator is aware of his own possible idiosyncrasies which can be checked out by more precise measurement using electronic lesion detectors and leaf area meters.

Problems will also arise from the fact that plants are very rarely attacked by one pathogen only. In such cases, separate measurements will need to be taken and, if the green leaf area remaining is also measured, then the percentage dead tissue not associated with the disease can also be calculated by subtraction.

Data from the visual assessment of plant tissue affected sometimes fails to correlate with the effect of the pathogen in reducing yield. There may be many reasons for this discrepancy, plant tolerance being one. Another reason suggested is that the visual symptoms do not correlate with the amount of tissue colonization, the latter being proposed as having more relevance to the physiology of yield. Precise techniques can now measure fungal biomass within the host tissues by either measuring fungal chitin (Ride and Drysdale, 1972) or ergosterol, another unique fungal constituent (Griffiths, Jones and Akers, 1985). Both methods involve sophisticated biochemical analyses and are thus only applicable to certain types of experimentation.

Severity measurements can be made by aerial infrared photography, the diseased crop tissue producing a different colour to that of healthy tissue. Other indirect measurements include the trapping or collection of spores by some means, the amount of spores produced in a crop having a highly significant epidemiological relevance in that a low spore production would imply a slow rate of disease increase.

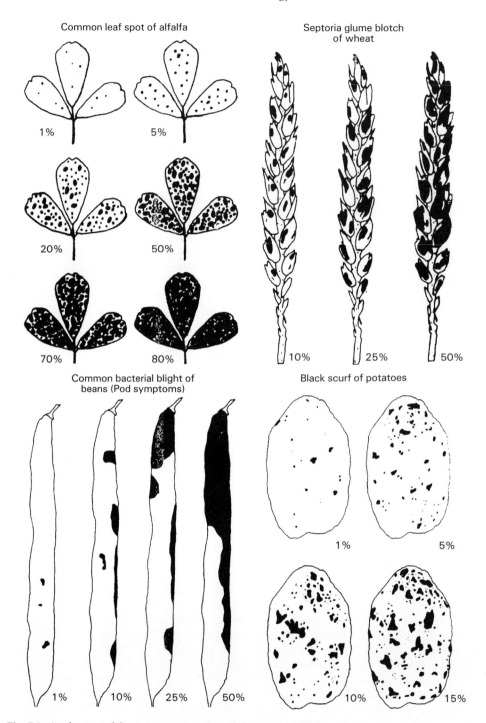

Fig. 7.2 *A selection of disease assessment keys* (after James, 1971)

Table 7.2 *Key for the Assessment of Potato Blight on the Haulm* (British Myco-logical Society, 1947)

Blight (%)	
0.0	Not seen on field.
0.1	Only a few plants affected here and there; up to 1 or 2 spots in 12 yards radius.
1.0	Up to 10 spots per plant, or general light spotting.
5.0	About 50 spots per plant or up to 1 leaflet in 10 attacked.
25.0	Nearly every leaflet with lesions, plants still retaining normal form: field may smell of blight, but looks green though every plant affected.
50.0	Every plant affected and about half of leaf area destroyed by blight: field looks green flecked with brown.
75.0	About three-quarters of leaf area destroyed by blight: field looks neither predominantly brown or green. In some varieties the young-est leaves escape infection so that green is more conspicuous than in varieties like King Edward, which commonly shows severe shoot infection.
95.0	Only a few leaves left green, but stems green.

Note: In the earlier stages of a blight epidemic, parts of the field sometimes show more advanced decay than the rest, and this is often associated with the primary foci of the disease. Records may then be made as, say, 1 + pf 25, where pf 25 means 25% in the area of the primary foci.

Most severity keys utilize a percentage scale where the grades are differentiated from each other in an exponential fashion. In this way, the keys more closely reflect the manner in which **compound interest** (see Chapter 8) or quickly multiplying diseases develop in the field. The potato blight key produced by the British Mycological Society (Anon, 1947) is the classical example of this type (Table 7.2).

The potato blight key is both numerical and descriptive. Some people find it easier to assess disease if pictorial keys are devised. The apple scab key (Croxall, Gwynne and Jenkins, 1952) incorporates pictures of both the leaf and the fruit stages of infection (Figure 7.3).

The assessment of root diseases has not yet been developed to such a precise art. It requires destructive sampling which obviously limits its use. A 0–5 scale for disease severity has been devised for the take-all pathogen (*Gaeumannomyces graminis*). It requires both destructive sampling and the longitudinal slicing of the roots to estimate the degree of vascular discoloration (Deacon, 1975). The assessment of cereal eyespot (*Pseudo-cercosporella herpotrichoides*) (Scott and Hollins, 1974) requires only visual observation, the tillers being taken at random and assigned to one of the described classes (Table 7.3).

A disease index can then be calculated from the formula:

$$\text{Disease index} = \frac{1(\text{tillers in class 1}) + 2(\text{tillers in class 2}) + 3(\text{tillers in class 3})}{(\text{total tillers in sample})} \times \frac{100}{3}$$

Sampling Methods

The research worker is not usually presented with sampling problems as he is likely to be dealing with individual plants or, at most, experimental plots. On a farm crop scale, for

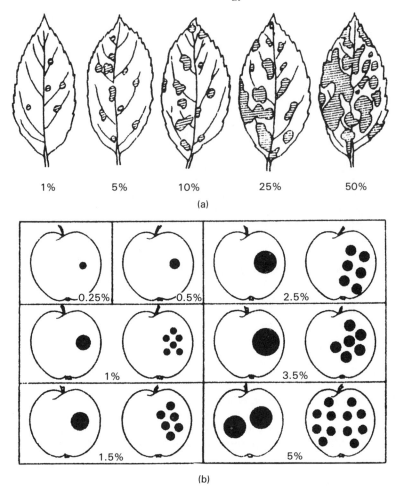

(a)

(b)

Fig. 7.3 *Standard diagrams for assessment of apple scab on leaves and fruit* (after Croxall *et al.*, 1952) © Crown Copyright, 1986.

Table 7.3 *An Assessment Key for Cereal Eyespot* (Scott and Hollins, 1974)

Symbol	Host: parasite reaction
0	Uninfected
1	Slight eyespot (one or more small lesions occupying in total less than half the circumference of the stem).
2	Moderate eyespot (one or more lesions occupying at least half the circumference of the stem).
3	Severe eyespot (stem completely girdled by lesions—tissue softened so that lodging would readily occur)

the purpose of disease surveys or for similar reasons, it will only be possible to assess disease on a sample of plants. The size of the sample will vary according to the precision required and the area of the crop being assessed. It is important that the samples are taken objectively and at random and that they are representative.

A common method is to sample along a predetermined line, possibly a diagonal across the field although, if samples are taken along a line the shape of the capital letter M, then more of the area will be covered. The distribution of disease throughout the crop will influence the sampling procedure, an even distribution requiring fewer samples than when disease occurs patchily. The actual numbers sampled can only be decided on the basis of previous experience and/or in relation to the statistical standard deviation of the samples, a reflection of the sample variation. In a field situation, a good indication of the incidence and severity of disease can be obtained by sampling at least fifty individual plants. In very extensive cropping, where large tracts are sown to one crop, it will be necessary to repeat this sampling in proportion to the crop area especially if there are obvious localized differences in soil type, drainage or topography.

Sampling may be destructive or non-destructive. The assessment of root diseases is of necessity destructive, it is easily facilitated in pot experiments but causes obvious labour problems in the field. Disease assessment, with the objective of calculating rate increases, can be carried out by random sampling on separate occasions or by repeated assessment of the same plants, initially selected at random, marked for easy identification by coloured tags, canes or aerosol sprays.

Timing of Assessment

Many assessments will be made for purposes of comparison or surveys. Assessment for the calculation of rate increase in disease is done at intervals. **Simple interest** (see Chapter 8) types of disease need only to be assessed on one occasion, but if the assessment data are to be used to estimate crop loss then the timing is crucial. With cereals, the timing of assessment should be made at a time when disease would most affect the physiological function of grain filling. For the purpose of comparison, growth stage keys have been devised which are now used internationally. Growth stages were first depicted graphically on the Feekes scale (Figure 7.4) (Large, 1954). With the advent of computerization, a decimal key has largely replaced the Feekes scale (Zadoks, Chang and Konzak, 1974) and has been illustrated (Figure 7.5) by Tottman, Makepeace and Broad (1977) (published by BASF (UK) Ltd). Other keys have been published for other crops including field beans and oil seed rape.

Crop Loss Appraisal

The most important single objective of disease assessment is crop loss appraisal both for yield and quality. Market pressure has resulted in the almost total disappearance of some quality reducing diseases of fruit, apple scab for example. With yield-reducing diseases, however, various attempts have been made to utilize disease assessment data for the estimation of yield loss. The conversion of disease data into crop losses is not an easy task. The classical work in this area was done by Large (1952, 1958) with potato blight. Working on the assumption that the bulking up of tubers ceases when 75% of the foliage (the **critical point**) is affected by blight, Large showed how the date on which this level of disease was reached could be related to the probable losses as estimated from potential yields if the crop was harvested at different times during the season. The two diagrams (Figure 7.6) indicate that the later the 75% level is reached, the lower the potential loss.

With cereals, as with many other crops, the effect of disease on yield can be related to the duration and severity of exposure to the disease. Large and Doling (1962) produced

Stage
1   One shoot (number of leaves can be added) = 'brairding'
2   Beginning of tillering
3   Tillers formed, leaves often twisted spirally. In some varieties of winter wheats, plants may be 'creeping' or prostrate
4   Beginning of the erection of the pseudo-stem, leaf sheaths beginning to lengthen
5   Pseudo-stem (formed by sheaths of leaves) strongly erected
6   First node of stem visible at base of shoot
7   Second node of stem formed, next-to-last leaf just visible
8   Last leaf visible, but still rolled up, spike beginning to swell
9   Ligule of last leaf just visible
10 Sheath of last leaf completely grown out, spike swollen not yet visible
10.1 First spikes just visible (awns just showing in barley, spike escaping through split of sheath in wheat or oats)
10.2 Quarter of heading process completed
10.3 Half of heading process completed
10.4 Three-quarters of heading process completed
10.5 All spikes out of sheath
10.5.1 Beginning of flowering (wheat)
10.5.2 Flowering complete to top of spike
10.5.3 Flowering over at base of spike
10.5.4 Flowering over, kernel watery ripe
11.1 Milky ripe
11.2 Mealy ripe, contents of kernel soft but dry
11.3 Kernel hard (difficult to divide by thumb-nail)
11.4 Ripe for cutting. Straw dead

© Crown Copyright

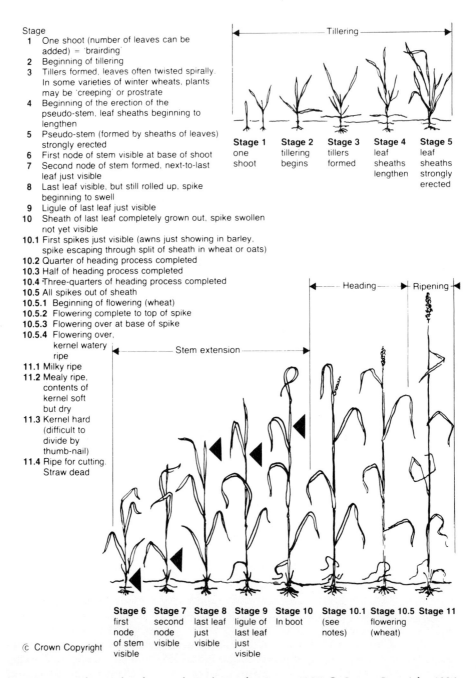

Fig. 7.4 *Growth stage key for cereals* (Feekes) (after Large, 1954) © Crown Copyright, 1986

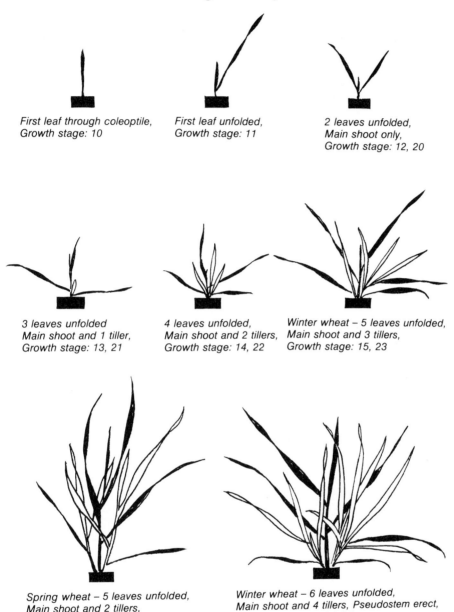

First leaf through coleoptile,
Growth stage: 10

First leaf unfolded,
Growth stage: 11

2 leaves unfolded,
Main shoot only,
Growth stage: 12, 20

3 leaves unfolded
Main shoot and 1 tiller,
Growth stage: 13, 21

4 leaves unfolded,
Main shoot and 2 tillers,
Growth stage: 14, 22

Winter wheat – 5 leaves unfolded,
Main shoot and 3 tillers,
Growth stage: 15, 23

Spring wheat – 5 leaves unfolded,
Main shoot and 2 tillers,
Growth stage: 15, 22

Winter wheat – 6 leaves unfolded,
Main shoot and 4 tillers, Pseudostem erect,
Growth stage: 16, 24, 30

Fig. 7.5 *A sample of illustrations of the decimal code for the growth of cereals* (after Zadoks, Chang & Konzak (1974), illustrated by Tottman, Makepeace and Broad (1977), published by BASF (UK) Ltd)

a formula which, they suggested, would give a reasonably accurate prediction of yield loss in barley due to powdery mildew infection: per cent loss = 2.5% $\sqrt{\text{mildew }\%}$ on flag leaf at GS 10.5 (Feekes scale). There are limitations to this formula as can be deduced from the fact that it only predicts a 25% loss if there is 100% mildew. Other workers have suggested

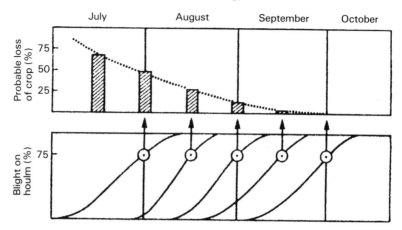

Fig. 7.6 *Graphs showing the estimated yield loss from epidemics of potato late blight in which the 75% disease level is reached at differing times during the season* (after Large, 1952; © Crown Copyright, 1986).

that a better relationship between diseases and yield is obtained if the area under the disease progress curve (see Chapter 8) is used as the predictor. These and other methods are described and discussed by James (1974), and by Teng, Close and Blackie (1979).

8

Plant Disease Epidemiology

Introduction

Until about the second half of this century, the study of plant disease mostly involved observations on individual plants. The information so obtained has been of great academic and, often, practical value but it was lacking in many respects. Diseases may be important on single plants grown in isolation but by far the most damaging effects occur when the plants are grown in a crop situation. Epidemiology is the study of disease development in populations of plants. In other words, the epidemiologist studies the progress of disease from its commencement, the inoculum source, through its rapid, exponentially increasing phase until it affects a high proportion of the population.

The epidemiologist concerns himself with all aspects of the disease triangle from the resistance of the host to the virulence and aggressiveness of the pathogen, the dispersal mechanism and the infection process and the interaction of each of these factors with the environment. Most significantly, epidemiology has introduced another dimension into the science of plant pathology for what was once an almost exclusively descriptive subject has now become analytical, what was once a qualitative subject has become quantitative.

Quite clearly, the study of an epidemic demands an accurate method of measuring disease. This may be achieved in many ways and the reader is referred back to the techniques of disease assessment already discussed in Chapter 7. When disease is measured over a period of time and plotted as a graph or **disease progress curve,** in many instances a sigmoid curve will result with the characteristic slow initial phase being followed by the exponential or logarithmic phase, when disease is multiplying at its fastest rate, ending up with a flattening out of the curve when, although there will be an abundance of pathogen propagules being produced which could initiate new infections, the target area of healthy tissue has almost disappeared (Figure 8.1).

The sigmoid disease progress curve will vary according to the nature of the host plant, the pathogen and the environment, the most common variation reflecting different rates of infection and disease development. The curve may also be displaced in time following some treatment that causes a delay in the epidemic.

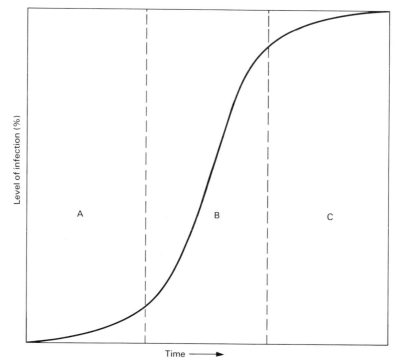

Fig. 8.1   *An idealized disease progress curve.*
  A—epidemic rate slow, low levels of inoculum, abundant host tissue
  B—epidemic increasing rapidly (log phase)
  C—epidemic rate decreases, abundant inoculum, little host tissue remaining

Inoculum Sources

Plant disease epidemics are caused, almost without exception, by the movement of the pathogen propagules, spores, bacterial cells, virus particles etc., from an infected plant to a healthy plant. There must be a starting point and this is termed the **inoculum source.** The identification of the source of inoculum is essential for the comprehensive study of epidemics and it can also provide the initial target for control measures. Inoculum sources can be either local or distant. The farmer may have some control over local sources such as infected plant debris or volunteer plants. He has little chance when the pathogen is dispersed over long distances; the spores of many rust pathogens can be carried over many hundreds of miles, for example.

   Inoculum source is synonymous with survival or perennation. However, the question is how the pathogen survives in the absence of the host species, during the winter period perhaps, or during the intercrop period in tropical and subtropical regions where more than one crop is grown during the year. In the latter case, the pathogen might have a relatively short period to survive or it may be the extremely hot, dry season which might necessitate some specialized dormant propagules if survival is to be accomplished.

Seed

A common source of inoculum is seed which may either be contaminated, as with the teliospores of bunt (*Tilletia caries*), or infected, as with mycelium of loose smut (*Ustilago nuda*). Contaminants offer no real problem to control by the use of disinfectants but infected seed demands a more stringent treatment, the use of a systemic chemical perhaps. In the case of *Septoria apiicola*, causing celery leaf spot, the infection is in the form of mycelium and the asexual fruiting bodies, the pycnidia, in the seed-coat but not the embryo. As the seed germinates, the pycnidiospores that exude from the pycnidia in the surrounding moisture can be spread to the emerging cotyledons and eventually, after the appearance of the seedling above ground, to the stems and foliage of the maturing plant. Some bacteria can also be seed-transmitted, *Pseudomonas phaseolicola* for example, but very few viruses survive the intercrop period in this way. The primary source of late blight (*Phytophthora infestans*) inoculum is the planting of infected tubers. Whilst the tubers are not true seed, they may be considered as such in this context.

Debris

Debris, or more precisely infected plant residues, are a common inoculum source and many epidemics are initiated from this origin. Infected wheat and barley leaves are known to be a major inoculum hazard in the glume blotch (*Septoria nodorum*), powdery mildew (*Erysiphe graminis*) and take-all (*Gaeumannomyces graminis*) diseases. Epidemics of apple scab (*Venturia inaequalis*) are initiated primarily from ascospores produced in perithecia embedded in infected leaves which overwinter on the orchard floor.

Volunteer Plants

Volunteer plants are also carriers of many rust pathogens, *Puccinia striiformis* for example, and infected potato tubers left in the ground can become important inoculum foci for the late blight disease (*Phytophthora infestans*) in the next season. In the latter disease, tubers discarded at harvest or during winter storage should be destroyed for, if they are left around in 'cull piles', they too provide a primary inoculum source as lesions on emerging sprouts sporulate the following season.

Cankers

Many bacterial pathogens overwinter in cankers on the bark of their tree hosts. The fire-blight of pears bacterium (*Erwinia amylovora*) provides an excellent example with its 'holdover' cankers, and the apple canker fungus (*Nectria galligena*) can also be found overwintering in this manner.

Alternative Hosts

Inoculum sources can also be identified away from the economic host species. The aecidiospores of *Puccinia graminis* are produced on the secondary or alternate host, the barberry (*Berberis vulgaris*) and the currant rust pathogen, *Cronartium ribicola*, has as its secondary host the five-needled pine. As both are economic crops, they can both be considered as alternative host species for this pathogen. Pathogens can also survive on related, alternative hosts, often weed species. The club root fungus (*Plasmodiophora brassicae*) can be found on the roots of cruciferous weeds such as charlock and shepherd's purse if these are allowed to grow in contaminated soil. Similarly, a few red clover plants can act as sources for inoculum for the pea mosaic virus.

Many fungi and bacteria have the ability to survive in the absence of the living host but, to succeed, they have to produce specialized propagules which can withstand extremes of temperature and desiccation. The resting spores of *Plasmodiophora brassicae* can remain viable for up to seven years, the oospores of *Pythium* and *Phytophthora* spp. have varying longevity from one year to several years with the resting spores of potato wart disease (*Synchytrium endobioticum*) surviving for up to thirty years. Other fungi produce aggregations of hyphae which can live in a dormant state in the absence of a living host. The sclerotia of such pathogens as *Sclerotinia* spp. and *Sclerotium cepivorum*, the onion white rot fungus, are good examples of soil-borne sclerotia, as are the sclerotia or ergots of *Claviceps purpurea* which can also contaminate the harvested cereal grain.

Dispersal of Inoculum

Unless inoculum is produced at the site at which the host plant is growing, it will have to be transported to that site for infection to take place. The transportation of some soil-borne pathogens poses more problems than the light, mass-produced air-borne propagules or the wet, perhaps sticky propagules of a great number of economically important pathogens. At one extreme, the root-invading pathogen *Gaeumannomyces graminis* survives the intercrop period on stubble debris and there is very little dispersal. Infection will only occur if a new host plant is sown where its roots will come into contact with the infected debris. At the other extreme, it has been suggested that spores of the coffee rust pathogen, *Hemilea vastatrix*, were blown across the South Atlantic to the East coast of Brazil. This can never be positively confirmed, but the elegant spore-trapping experiments of Hirst and Hurst (1967) from an aircraft over the North Sea certainly indicate that long-distance dispersal is common. The study of spore dispersal is essential to epidemiological investigations and, ultimately, can provide the weak link at which to direct a control measure. The rust pathogen *Cronartium ribicola* has two hosts, *Ribes* spp. such as currants and gooseberries, and the five-needle pine. With the longevity of the basidiospores being extremely short, about as long as it takes for them to be blown 300 metres, the pines can be protected from infection by ensuring that there are no *Ribes* bushes within that distance of the tree crop.

Wind

The main agents of inoculum dispersal are wind, water, insects, animals and man. Of these, wind is by far the most important, being involved in the dispersal of rusts, mildews, late blight of potato, apple scab and brown rot of apples and pears and many other pathogens of world-wide importance. Air-borne spores may be trapped and identified using a variety of techniques ranging from the very simple Vaseline-treated microscope slide to the more sophisticated volumetric spore traps with which it is also possible to quantify the concentration of spores present in the air at the sampling site (Plate 8.1).

Water

Wind is also involved in the dispersal of rain-splashed pathogens, blowing away the fine spore-carrying water droplets although the distances involved may not be great, 1 metre for example in the spread of *Septoria nodorum* in a wheat crop (Griffiths and Hann, 1976). Some fungal spores appear to be dispersed almost exclusively by rain, the coffee berry pathogen (*Colletotrichum coffeanum*) for example. Again, simple spore traps can be devised from the combination of glass or plastic funnels and containers such as medical flats or milk bottles. Many bacteria are also spread by rain, not only by the splash mechanism but

Plate 8.1 A seven-day recording volumetric spore trap for battery or mains operation (Photo: Burkard Manufacturing Co. Ltd)

also by runoff or dripping rain water or water from irrigation sprinklers. The spread of the fire-blight pathogen (*Erwinia amylovora*) is effected by rain splashing on the 'holdover' cankers, and many bacterial leaf spots, such as angular leaf spot of cotton (*Xanthomonas malvacearum*) are also spread by this agency.

Insects

The spread of viruses by insect vectors has already been discussed in some detail in Chapter 3. Such insects include aphids, leafhoppers, thrips, whiteflies, beetles and mealy bugs. In quantitative terms, the spread of insect-borne virus diseases will be determined by the numbers and activity of the vector. The environmental conditions prevailing at any given time must be taken into account when estimating potential disease levels and can form the basis for the forecasting of disease levels. Many fungi are also spread by insects, and probably the example which has had the greatest impact on man and his surroundings this century is the spread of the Dutch elm disease pathogen (*Ceratocystis ulmi*) by the bark beetle (*Scolytus* spp.).

Animals and Man

Dispersal by other animals including mammals, birds, nematodes and, of course, Man has been well documented. Examples include such diseases as the oak wilt fungus (*Ceratocystis fagacearum*) which may be spread by the activity of squirrels, the chestnut-blight disease (*Endothia parasitica*) spread partly by birds such as woodpeckers, the many nematode-transmitted viruses and the many pathogens dispersed by the activity of Man and his agricultural practices. Man has been implicated in the long distance dispersal of many pathogens through the transport of infected vegetative propagating material such as bulbs, corms, seeds and nursery stock. He may also be involved in the spread of soil-borne pathogens in the soil on his boots or on the tractor wheels. The feeding of animals with root crops such as turnips or swedes infected with *Plasmodiophora brassicae* is a sure way of moving the long-lived resting spores from field to field via dung. Labour-intensive crops, especially those grown under glass, are all candidates for Man's indiscretions in this context. The repeated picking of cucumbers throughout the season provides the dispersal mechanism for the angular leaf spot pathogen, *Pseudomonas lachrymans*, which is easily picked up in the moisture and spread on the contact of human hands and the plant surface.

The Environment

Without conditions conducive to disease development, epidemics would never occur. The disease triangle describes the tripartite nature of the epidemic and the interactions which occur. The environment might well be considered as the controlling influence, affecting the magnitude of most epidemics, their effects beginning with the preinfection stages of inoculum survival and extending to spore germination, penetration, colonization, sporulation and dispersal.

The environmental factors which should be considered in this context are temperature, rainfall, air humidity, wind and duration and intensity of light. These factors can all be investigated as separate entities, but it is important to remember that they all interact with each other.

Temperature

Temperature is the controlling force for all biological processes and it is easy to extrapolate this concept to plant diseases. In the main, within a wide range of commonly occurring temperatures, 5–30 °C for example, disease development will increase with increasing temperature. This is especially so with bacterial diseases although there are the inevitable exceptions with diseases like halo blight of oats (*Pseudomonas coronofaciens*) being favoured by cool temperatures. Similarly, with fungal pathogens, there are exceptions like the snow mould pathogen of wheat (*Micronectriella nivalis*) which is particularly damaging on winter wheat under a heavy snow cover. Temperature, of course, interacts with humidity and surface moisture so high temperatures may cause a reduction in spore germination in many pathogens.

The effect of temperature may also be seen on the degree of sporulation on a plant and this, in turn, will greatly influence the rate of epidemic development. Jeger, Griffiths and Jones (1983) published some very interesting results on seasonal variation in the components of partial resistance in wheat to *Septoria nodorum*. In March, April and May, for example, the ratio of sporing lesions on cultivar Maris Ranger to cultivar Maris Huntsman was 10:1, but by June the ratio had changed to 3:1, indicating not only a temperature effect but also a cultivar:temperature interaction. In the same paper, results showed that both cultivars increased their sporulation rates as temperatures increased during these four months.

Many other effects could be discussed. There is ample evidence of an inverse relationship between latent period and temperature in a number of diseases. In *Puccinia graminis* for example, latent period decreased from 22 to 5 days as the temperature increased from 5 to 24 °C (Stakman and Harrar, 1957).

Light

Many pathogens have been shown to respond to light. In particular, some fungi are photosporogenic, as with the example of *Septoria nodorum* (Cooke and Jones, 1970) which can be induced to sporulate profusely if incubated under 'black light' or, more correctly, near ultra violet (NUV) (see Plate 7.1). Much work has been carried out on sporogenicity by Leach (1962). Maddison and Manners (1972, 1973) have shown that the germination of urediospores of *Puccinia* spp. varies greatly with spore type, the presence of pigments in spores helping to absorb harmful rays. However, light does not appear to have very much effect on the epidemic once infection has occurred.

Humidity

Almost all fungal spores can be stimulated when they are placed in a film of distilled water. Free water is not always necessary and a relative humidity exceeding 99% will also induce germination. The exceptions to this are the powdery mildews whose conidia can germinate at 0% relative humidity due to their exceptionally high water content (70% compared with about 10% in most fungi). However, even the powdery mildews germinate best at relatively high humidities and optimum germ-tube growth occurs at about 98% relative humidity.

Use is made of the knowledge that most fungi require **moisture** and high humidities inasmuch as most artificial inoculation techniques involve the spraying of a spore suspension in water to provide the free surface moisture, followed by the placing of the inoculated plant in a humidity chamber, often only a plastic bag, to ensure a period of high humidity.

Soil moisture is, of course, essential for both infection and movement within the soil environment of many soil-borne pathogens. The zoosporic fungi in particular, the club root pathogen *Plasmodiophora brassicae* and many damping-off and root invading pathogens of *Pythium* and *Phytophthora* spp. are good examples, and require adequate soil moisture for successful infection and movement of zoospores.

Other Factors

There are numerous other factors that can and do affect stages of the disease cycle, from the pH and nutrient status of the soil, its compaction and aeration to the wind that can affect the physical dispersal of spores, infected plant debris or the activity of insects which are potential carriers of virus particles. Many aspects of spore dispersal have been discussed by Gregory (1973), and many textbooks cover environmental effects in terms of epidemiology (Agrios, 1969; Tarr, 1972).

Quantification and Interpretation of Epidemics

The precise measurement of disease is essential if epidemic progress is to be plotted over a time-scale and if comparisons between epidemics are to be made. The subject of disease assessment has already been considered (Chapter 7). Increase in disease can be measured either as **incidence**, where the number of infected individual plants will be recorded, or as

severity, where the degree of infection or tissue damage, visual disease symptoms, is recorded.

The plotting of successive disease measurements during an epidemic will facilitate its interpretation in terms of time-scale, rate of increase and maximum disease levels. There will inevitably be variation as regards these factors and the causes of such variation can be attributed to one, two or all three components of the disease triangle. The pathogen might vary in terms of the virulence genes present and the amount of initial inoculum, the host may vary in terms of its resistance to the pathogen virulences and, of course, the environment will be superimposed as has been discussed already in this chapter.

Not all diseases multiply in the same way. At one extreme, there are those diseases which start usually from soil-borne inoculum and infect any hosts with which they come into contact. There is only one cycle of disease per growing season, that is, only one generation of the pathogen from infection to sporulation. These are called the **monocyclic** diseases and their multiplication has been likened to **simple interest** in banking terms where the capital is increased only once a year. Simple interest diseases are quite common and good examples are caused by the soil-borne pathogens *Fusarium*, *Verticillium* or *Plasmodiophora* where the initial inoculum is added to once a year on the release of spores from the decaying, infected host plants.

In contrast, some diseases multiply rapidly, going through several cycles during the season. These are called the **polycyclic** diseases and their multiplication can be compared with **compound interest**. These financial analogies were propounded by Van der Plank (1963) in his textbook on epidemiology and he must be given the credit for most of the current interest in the subject.

When the development of a 'compound interest' disease is plotted, the graph will be sigmoid with the characteristic slow start, the lag phase, leading into the exponential or logarithmic phase and then flattening out as the availability of host tissue rapidly diminishes due to its depletion by disease (Figure 8.1). The 'simple interest' type of disease would not give a sigmoid curve during one season as increases in the pathogen population would only occur once every year. If plotted over several seasons, however, the progress curve would assume a sigmoid nature although the exponential slope would be erratic in its increase rather than smooth, reflecting differences in seasonal multiplication.

The analogy with the growth of invested capital is very appropriate as the formula for calculating compound interest is also applied to the early exponential part of the progress curve:

$$x = x_0 e^{rt}$$

where
    x = the amount of disease at a given time
    x_0 = the amount of disease at the start of the epidemic
    e = mathematical constant
    r = rate of increase of disease
    t = time

As the disease progresses and the amount of tissue available for infection becomes depleted in an accelerating manner, a correction factor must be employed to modify the original equation. Thus, as the epidemic progresses, its rate is governed both by the amount of disease present, x, and by the proportion of susceptible tissue left $(1 - x)$.

So, where disease has been assessed at intervals during the epidemic, it is possible to calculate the one unknown in the equation, that is the rate, r, by calculating the average increase of disease between two dates of assessment, t_1 and t_2. The equation for this is shown below (2) and incorporates a log transformation, the purpose of which is to straighten out the sigmoid curve:

$$r = \frac{1}{t_2 - t_1} \log_e \frac{x_2}{x_1} \qquad (2)$$

To illustrate the use of the equation, if *Septoria nodorum* on wheat was assessed on 5 June and 15 June and found to be 0.5% and 3.5% respectively, we can substitute in the equation $x_1 = 0.005$, $x_2 = 0.035$ and $t_2 - t_1 = 10$ days.

Thus,
$$r = \frac{1}{10} \log_e \frac{0.035}{0.005}$$
$$= 0.194 \text{ unit increase in disease per day.}$$

This calculation gives a value for the rate of increase in disease at the start of the epidemic but, when about 5% of the host tissue has been infected, the correction factor must be included. The formula for the rest of the exponential phase would be:

$$r = \frac{1}{t_2 - t_1} \log_e \frac{x_2(1 - x_1)}{x_1(1 - x_2)} \qquad (3)$$

More of the mathematical background to these formulae with example exercises can be found in Van der Plank (1963). There is some evidence that the above equations do not give precise fits to epidemics of all compound interest diseases but they do serve as invaluable starting points for epidemic comparisons and interpretations.

From a practical point of view, the compound interest equation (1) gives clear directives as to the alternative approaches to disease control. The equation highlights the three main components x, r, and t. Any control treatment which alters x_0, r or t—that is, the right hand side of the equation—will obviously produce changes in x on the left-hand side of the equation. In other words, the pathologist is invited to produce control treatments that will reduce x_0, r or t as these must lead to reductions in x, the amount of disease.

The principles of disease management are discussed in Chapter 9. Methods of reducing x_0, the initial amount of inoculum, are either concerned with race-specific resistance, which eliminates certain races from the pathogen population, or with sanitation or crop hygiene treatments which can vary from the use of seed disinfectants to the destruction of infected plant debris. Methods of reducing x_0 have been practised for centuries and are often classified as cultural. Probably the oldest of these methods is crop rotation which can be very effective in eliminating soil-borne pathogens. The rotations may be short or long, the result the same, a reduction in the surviving inoculum, a reduction in x_0.

The effect of reducing x_0 can be predicted by use of the compound interest equation (1). If we consider an epidemic during the exponential phase then:

$$x = x_0 e^{rt} \qquad \text{with no sanitation}$$

and
$$x_s = x_{os} e^{rt} \qquad \text{when sanitation has been applied}$$

Division gives: $\dfrac{x}{x_s} = \dfrac{x_0}{x_{os}}$

From this latter equation two important points emerge. Firstly, x_0/x_{os} reflects the amount of sanitation achieved, in other words, it is the **sanitation ratio**. Secondly, the effect of sanitation, that is, the proportion of the final amounts of disease in the crop with and without sanitation, x_0/x_{os} remains constant throughout the exponential phase. This interpretation will be better appreciated by substituting some figures. If 90% of the initial inoculum x_o is eliminated by a sanitary measure the sanitation ratio can be expressed as,

$$\frac{x_o}{x_{os}} = \frac{100\%}{100\% - 90\%} = 10\%$$

Thus the final proportions of disease will be $x/x_s = 10\%$.

So there will be, at the final assessment, only 10% of the amount of disease that there would have been without sanitation and this ratio would be constant throughout the exponential phase of the epidemic.

We can look at the effect of sanitation in another way. As the effect of sanitation remains constant during the rapidly multiplying stage of the disease, we can evaluate the effect of sanitation in terms of the delay it causes in the epidemic, a **delay time** which will also be constant during this period. The delay time, which can be expressed as Δt, can again be calculated from our original equation with modifications to incorporate sanitation. Thus:

$$x_o = x_{os} e^{rt}$$

and

$$\Delta t = \frac{1}{r} \log_e \frac{x_o}{x_{os}}$$

If we substitute in the equation with example measurements:

$$r = 0.3$$

and

$$\frac{x_o}{x_{os}} = 10$$

Transforming from $\log_e$ to $\log 10$

then

$$\Delta t = \frac{2.3}{0.3} \log_{10} 10$$

$$= 7.7 \text{ days}$$

This is a very informative calculation and one which can be plotted (Figure 8.2) where it can be seen that the delay is constant throughout the exponential phase. In diseases where potential yield is increasing during the period of the epidemic, the delay will be translated into an increase in yield as the treated crop remains relatively more healthy and hence productive during this time. This is best seen by comparing the potato blight (*Phytophthora infestans*) disease progress curves for the treated (with sanitation) and untreated (no sanitation) crops with the graph for the likely yield of tubers in a healthy crop if harvested at any time during the season (see Figure 7.6, p. 88). If the 75% blight level is taken as a **critical point** in this disease after which no further yield increase may be expected, then the delay of 7.7 days can be transposed on to the yield graph to predict the likely yield increase due to the sanitary measure.

The rate of disease development can best be altered by the use of host resistance or chemicals. Resistance of the 'horizontal' category, in which various components are invoked in an additive manner to delay and reduce the infection cycle, will result in a reduction in the rate, r. Any resistance comprising a lowering of the infection frequency, a lengthening of the latent period or a reduction in sporulation will be rate-reducing. Chemicals, of the protectant rather than eradicant category, by killing large numbers of spores as they land on the treated host surface or by slowing down fungal growth will also have the same effect.

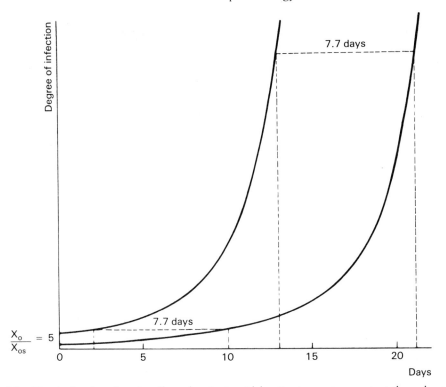

Fig. 8.2 *Graph showing that the effect of sanitation (delay time) remains constant throughout the logarithmic phase of disease development. (after Van der Plank, 1963)*

Given the right environmental conditions, serious epidemics can arise from very small amounts of initial inoculum so, although we have just considered the benefits from sanitation, it must be emphasized that reductions in the rate of spread r and the time of spread t will often be more beneficial. The level of infection at any given time in a crop gives an indication of the potential damage that might occur as the epidemic progresses. However, such single assessments can be very misleading if the rate of disease increase is ignored. Whitney (1976) gives a very good example of the effect of rate in comparing two epidemics, the first where the pathogen is very aggressive and multiplies very rapidly from a low level, the second, where the pathogen is less aggressive and multiplies slowly but from a larger amount of initial inoculum. Figure 8.3 shows the outcome of both epidemics with the disease levels at the time of assessment identical. With the rate in epidemic ① 25% greater than in epidemic ②, by harvest the final amounts of disease are very different and the expected damage would be much more with the higher rate epidemic even though it started from a lower initial inoculum.

Changes in the time factor will also affect the course and importance of the epidemic. If disease is confined to a short period only, and that period is relatively unimportant in terms of the effect upon yield, then each unit of time will be of less significance than at a more physiologically important phase in the plant's development. Similarly, changes in the latent period—that is, the time between infection and sporulation, will also affect the rate of disease development, an increase in latent period reducing the number of disease cycles possible in a given period of time.

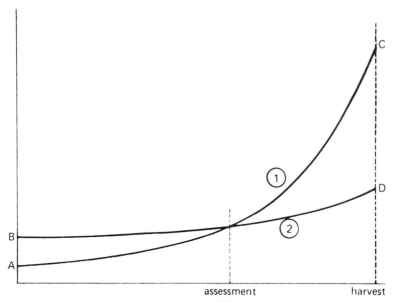

Fig. 8.3 *Graph showing the importance of the rate of development of a disease and how this cannot be assessed by a single observation. Observation at the time indicated would give the impression that epidemics 1 and 2 were of equal magnitude, while observations before this time would indicate outbreak 2 to be the more serious*

$A = \frac{1}{3} B$
$C = 2.5 \, D$

Rate 1 is 25% greater than rate 2 (after Whitney, 1976)

Prediction and Modelling of Epidemics

The ability to forecast the arrival of inoculum or the onset of an epidemic has obvious benefits both to the pathologist in his study of plant diseases and to the farmer who is afforded time to prepare his control treatments and to time their application more efficiently. Mostly, forecasts will be of a short-term nature but, if the forecast is long-term, it would also allow a change in the cultivar grown from perhaps one which is a high-yielding but susceptible, to one which is more resistant and will yield well in the presence of the pathogen.

In the development of forecasting systems, comprehensive data must be available of the crop and its development through the season, the pathogen and of the interactive effects of the environment on disease development. Not all diseases require an elaborate forecasting system. To warrant research in this respect, the disease must be economically important, it must be variable both in its time of arrival and the damage caused and, quite obviously, adequate control measures must be available to combat the particular disease.

The amount of initial inoculum, dispersal factors, numbers and activity of vectors and prevailing environmental factors all contribute to oscillations in the incidence, severity and onset of disease. In some diseases, apple scab (*Venturia inaequalis*) for example, a few dominant factors (such as leaf wetness and temperature) in conjunction with the presence of inoculum form the basis for efficient forecasting methods, and similar factors are used in many other forecasting systems.

With such prerequisites for a forecasting system, the early detection of a small population of spore inoculum or the initial infections in a crop would seem essential. In some soil-borne diseases, estimates of inoculum in soil samples can be made as can estimates of inoculum on certain planting material, such as the incidence of skin spot-lesions (*Polyscytalum pustulans*) on potato tubers. However, in general, the measurement of pathogen populations is a limiting factor to the forecasting of many diseases. With air-borne inoculum, several types of spore-trap have been developed to monitor the pathogen population. A volumetric spore-trap (Plate 8.1) may be used not only to detect the initial release of spores, as in the ascospore population in apple scab, but also population size may be determined. Severe disease is always more likely in the presence of large spore populations. A good example of this can be seen in the forecasting of early blight of potato (*Alternaria solani*). As the season progresses, inoculum will be produced on plant debris in the first instance and later, as the plants become more susceptible with maturity, large spore populations will be detected indicating the need for the application of an appropriate fungicide. Various techniques have been utilized in disease prediction and these are discussed below.

Modelling Epidemics

There are three types of models: (1) descriptive (2) predictive (3) conceptual. The **descriptive model** has already been discussed, the plotting of disease progress curves being the result. Mathematical equations are available to describe these curves but they have no biological meaning in themselves, the equation $x = x_o^{e^{rt}}$ being also the equation to describe compound interest of money.

The **predictive model** is based on either regression equations or differential equations. They are biologically more meaningful than descriptive models as they utilize data from disease assessment and from meteorological observations. The **conceptual model** should give very precise predictions as it incorporates the interactions of all the factors comprising the epidemic including spore production, dispersal, infection, host resistance and the environment. If all the measurements were accurate and all the interactions calculated precisely, the prediction of disease would be perfect. This is normally unattainable and only reasonable estimates made of certain parameters. The so-called 'disease simulator' is a development of the conceptual model, and the most famous of all is the simulator of Southern corn leaf blight (*Helminthosporium maydis*) produced by Waggoner, Horsfall and Lukens (1972) and called EPIMAY. In this, every minute detail of the disease cycle was studied, data collected and incorporated into the model. Even minor interactions of weather and spore production factors were included. EPIMAY was produced in response to the great epidemic of Southern corn leaf blight which occurred in 1970.

The epidemimetic model or simulator was ready by June 1971 and the aim was for it to accurately mimic the epidemic, hence the adjective epidemimetic. It comprised three main components: (1) weather as observations; (2) the pathogen and all its responses to the weather; (3) the computer programme which, in essence, predicted how the pathogen would react to the current weather. The calculations indicate whether the epidemic will rise, fall or remain unchanged. So, at any time, given the necessary data on weather and current disease level, EPIMAY acts as an early warning system of likely changes in the epidemic.

EPIMAY was not the first simulator as EPIDEM had been developed in the same laboratory a few years earlier in 1969 for the simulation of epidemics of early blight of tomato and potato (*Alternaria solani*). Several others have since been developed such as BARSIM-I (Teng, Blackie and Close, 1980), for barley leaf rust (*Puccinia hordei*), and EPIPRE (Zadoks, 1981) which has achieved considerable patronage for the prediction of several cereal diseases in the Netherlands, its country of origin.

The examples of disease forecasts based on conceptual models are very few and, in the main, forecasting methods have been developed from regression equations or empirical relationships.

Regression Analysis

Where disease increases in proportion to changes in some readily measurable influencing environmental factors, a relationship may be calculated by regression analysis. The simplest form, **linear regression,** relies on disease being related linearly to one parameter. For example, if the spore inoculum of a soil-borne fungus such as *Plasmodiophora brassicae* is estimated by using plant infection tests in the greenhouse, then the results of such tests could be used to compare with actual field observations of disease incidence at a later date. In other words, if there was a good agreement between the infection tests and the actual disease measurements, then a prediction could be made of the likely level of disease in soil by extrapolating from the regression lines.

When actual spore numbers are counted, it might be assumed that there would be a quantitative relationship between the inoculum amount and the amount of disease which accrues after a given time. This relationship is never simple and the possibility of a threshold number of spores being necessary before infection can occur is one to be seriously considered. However, research has shown that with several pathogens, at low inoculum levels, the number of lesions was directly proportioned to the spore-dose and that the regression line passes through the origin. A similar result was reported by Van der Plank (1967) for *Synchytrium endobioticum* and, therefore, it is generally assumed that the amount of infection is proportional to the amount of inoculum. This is certainly true of many bacterial and virus diseases but, it should be pointed out, the plotting of such dosage response curves will normally give a sigmoid curve and the straight line will only be produced by a transformation, possibly a log–log transformation or the use of probits.

Diseases are the result of a complex of interacting factors and it is rarely possible to correlate disease level with a single factor. For example, the severity of a virus epidemic may be related to the number and activity of an insect vector and to factors of the environment which might affect insect survival in the winter and insect multiplication and spread the following spring. Such is the case with sugar beet yellow virus which is predominantly transmitted by the aphid, *Myzus persicae*. Watson (1966) found a very close relationship between the amount of disease in the years 1950–66 and the predicted values calculated from a *partial regression* analysis. The two components she used in this analysis were the number of freezing days in January, February and March and the mean weekly temperature in April. With such a good relationship, Watson suggested that a forecast of the likely level of disease in any year should be possible by the end of April. The same approach can be made when several influencing factors are measured. This is then based on a *multiple regression* analysis using the equation:

$$y = a + b_1 x_1 + b_2 x_2 \ldots b_n x_n$$

where $x_1, x_2 \ldots x_n$ are the independent variables and $b_1, b_2 \ldots b_n$ are the corresponding partial regression coefficients.

From the above it can be seen that the use of regression analysis can be used in the prediction of possible disease levels, the prediction being based on accumulated data from past epidemics. In contrast, the simulator incorporates known responses of the pathogen to any permutation of influencing factors and can take into consideration the resistance likely to be offered by the cultivar present and, by its programmed sensitivity to various components of the epidemic, spore germination, penetration, sporulation, for example, will predict the likely magnitude of the next step in the epidemic development.

Disease Forecasting: Examples

Reference has already been made to several predictive techniques. The simulators like EPIMAY and EPIDEM for specific diseases or the more general approach for pests and diseases with an input from the farmer himself in EPIPRE (Zadoks, 1981) are good examples of the integration of biology, decision-making and computer technology. There are many other examples of disease forecasting in use today, some very empirical and others using a combination of techniques.

Potato Blight   This disease has, in the past, been largely controlled by the application of protectant fungicides. The timing of application has been shown to be crucial in late blight of potatoes (Evans, Couzens and Griffiths, 1965) and this is best facilitated by an accurate forecasting system. This was recognized many years ago, and Beaumont (1947) devised a forecasting scheme based on the incidence of periods of high relative humidity and fairly high temperature. The **Beaumont periods** were of 48 hours duration when the temperature was more than 10 °C and the relative humidity more than 75%. The occurrence of such periods was broadcast in the United Kingdom weather programmes and, generally speaking, could be relied on to precede the blight epidemic. There were exceptions, however, due to the regional basis of the forecast which did not take into account local topography or the condition of the crop.

The Beaumont period provided a good start to blight forecasting, albeit empirically based. In an attempt to refine the forecast, the **Smith period** (Smith, 1961; Martin and Worthing, 1976) was introduced which comprised two consecutive 24 hour periods (midday to midday) in which the temperature was more than 10 °C and incorporating at least 11 hours on each day when the relative humidity exceeded 89%, in other words, periods of leaf wetness. These measurements could be taken within the crop and made the forecast more precise.

Recognizing that an even more precise pattern could be produced if local meteorological data was fed into a centrally based computer, Krause, Massie and Hyre (1975) devised the **Blitecast** forecasting system in the eastern states of the United States. Again, the readings of temperature, rainfall and humidity were taken within the crop and the subscribing farmer phoned in such data for the preceding week. The proponents of Blitecast would claim that much of its success was due to the fact that the computer-based advice would often be to delay spraying, hence saving the farmer perhaps one to three otherwise wasted insurance spray applications. A recent modification of the Blitecast system corrects the forecasts for cultivar resistance and the application of fungicides (Bruhn and Fry, 1981).

Apple Scab   A similar system has been devised for apple scab predictions with the addition of data on the presence of inoculum being the overriding factor. Spore-traps can be placed in the orchard to monitor the release of ascospores in the spring and early summer and, in conjunction with **Mills periods** (Mills and La Plante, 1954), which are combinations of temperature and leaf wetness for varying durations, accurate forecasts can be made which facilitate a more efficient approach to fungicide application (Table 8.1).

Table 8.1   Examples of Mills periods

Mean temperature (°C) over period	5.6	7.2	10	11.7	15
Hours of leaf wetness	30	20	14	11	10

Leaf wetness is measured using a 'surface wetness' recorder which is a simple device which substitutes the leaf by a polystyrene block, changes in the weight of which can be measured on a pen-recorder as surface moisture varies.

Again, slight modifications of the Mills period have been made in the Smith period (Smith, 1961) which utilize a threshold of 90% relative humidity following rain and hence eliminates the requirements for the surface-wetness recorder.

Other modifications have again utilized the computer's enormous calculating capacity. A good example of this is the apple scab predictor which was recently evaluated by Ellis, Madden and Wilson (1984) who showed that the scheduling of fungicides with curative activity gave excellent control when applied 96 hours after initiation of the predicted infection periods.

Cereal Diseases   Several of the more economically important cereal diseases are now predicted on the basis of meteorological data for the purpose of decision-making in terms of fungicide application. In the United Kingdom, the **high-risk periods** used for the prediction of barley powdery mildew epidemics (*Erysiphe graminis hordei*) (Polley and King, 1973) are based on:

(1) Day maximum temperature $>15\,°C$
(2) Day sunshine >5 hours
(3) Day rainfall $<1\,mm$
(4) Day run-of-wind $>246\,km$

A high-risk period comprises several **high-risk days** each of which satisfies the above four criteria. The high-risk period ends on the first day when none or only one of the criteria have been satisfied. Taken in conjunction with the age of the crop and the presence of mildew inoculum, efficient spray applications can be made.

The current recommendations for spraying wheat for the control of *Septoria nodorum* are based on (a) protecting the plant at its most critical time in terms of possible future reductions in grain yield due to disease, that is GS 8–10 (37–43 decimal scale); and (b) the occurrence of favourable conditions for infection:

(1) Rain on two out of three days totalling at least 10 mm with rain on the first day,
(2) Rain on three consecutive days totalling at least 5 mm,
(3) Rain on four consecutive days.

In addition, a relative humidity of at least 90% must be recorded on at least one day in each period.

With eyespot of wheat (*Pseudocercosporella herpotrichoides*) Ferhmann and Schrodter (1972) have defined the conditions favourable for infection as 4–13 °C and relative humidity $>80\%$. In the United Kingdom, if such conditions occur and disease is readily detectable in the crop at GS 5–6 (26–31, decimal scale) then a spray application is highly advisable.

The EPIPRE system (Zadoks, 1981) developed in The Netherlands is used for both pests and diseases of wheat. In 1977, only yellow rust (*Puccinia striiformis*) was included in the system. By 1985, the system had not only spread to other countries but also had expanded to include *Puccinia recondita*, *Erysiphe graminis*, *Septoria nodorum* and *Septoria stritici*. There can be no doubt that disease forecasting is here to stay. It may still be rather empirical and there may still be inaccurate forecasts but, with the ever-increasing input of biological data and the continued refinements to computer programs this aspect of disease management cannot fail to be improved.

9

Plant Disease Management

Introduction

In recent years, the term 'disease management' has tended to replace the more emphatic description 'disease control'. Ideally, the aim of all plant disease management systems is to effect complete control. This is rarely achieved and, in practical terms, disease management will be judged to be successful if it can produce significant economic returns. It should also be emphasized that management strategies are primarily intended to save crops rather than individual plants. In addition, with individual management treatments rarely being completely effective, it is recommended that an integrated programme of treatments and cultural practices be followed, the methods employed producing complementary, additive and beneficial effects on crop yield. Such programmes may require repetition over several seasons before disease is reduced to negligible and acceptable levels.

Modern disease management requires a detailed understanding of all aspects of crop production; economic, environmental, cultural, genetic and epidemiological information providing the input data upon which the management decisions are made. In the present vogue of producing 'blueprints' for crop production, it is essential to recognize disease management as an integral component, to appreciate the magnitude of possible damage caused by disease, and to be aware of and select the appropriate programme of management strategies.

Almost all disease management strategies are aimed at preventing infection in the first instance, or the spread of the pathogen if this fails, as it often does. Truly curative measures are rare although certain heat treatments and the use of systemic chemicals may be categorized as such. The choice of treatment will depend very much on epidemiological information, often utilized to produce forecasting systems. The timing of the treatment may be chosen on the basis of differences in pathogen vulnerability during its life-cycle or, perhaps, by the necessity to protect highly susceptible periods in the crop's development. It is necessary to elucidate the overwintering pattern of the pathogen, or how it exists during an intercrop period, and to be fully acquainted with its mode of dispersal, infection and any interactions with environmental factors.

Without such basic information, disease management is bound to be imprecise, based on arbitrary decisions and often leading to a 'blanket approach', a too generous and badly timed spraying programme perhaps, which is inevitably wasteful, inefficient and likely also to be uneconomic.

With the tremendous diversity of pathogens and disease situations, it follows that disease management techniques, of necessity, will be equally varied. The following list groups the techniques according to their underlying objectives:

(a) exclusion of the pathogen;
(b) eradication of the pathogen or reduction of the inoculum;
(c) protection of the host;
(d) adjustment of the environment;
(e) use of plant resistance.

However, disease management techniques should be applied only if disease is adjudged important or predicted to be so and even the casual observer will appreciate that the incidence and severity of plant diseases vary considerably from season to season and for a variety of reasons. Predicting the importance of disease is a most difficult task in itself, but its success will largely depend on the prediction or 'forecasting' of probable outbreaks or increases in intensity of disease. Disease forecasting is, therefore, an integral and essential prerequisite in disease management decision-making.

Exclusion

Even in recent times, the devastating impact of certain introduced pests and diseases emphasize the importance of excluding potentially damaging pathogens from areas where they do not exist. In many countries with a relatively recent history of agricultural development, for example New Zealand, many of the pathogens of major economic importance are not original residents of that country but have been introduced, by some means or other, during the last century. Such introduced pathogens will encounter hosts which have developed no resistance against their presence and epidemics can become severe and of common occurrence.

Successful exclusion methods depend, in the main, on legislation which may be used to enforce prohibitions, quarantines or inspections. Such regulatory methods may also incorporate certification schemes linked to purity and viability of commercial seed.

The overall objective of quarantines and inspection is one of exclusion from a locality or region but mostly from a country. However, it is doubtful whether any such measure will be permanently effective. Total prohibition is sometimes ordered and, in several instances, has been successful. The ban on the import to the United Kingdom of seed potatoes from countries like the United States, where bacterial ring rot (*Corynebacterium sepedonicum*) is endemic is one such example, and the restrictions on lettuce seed imports (lettuce mosaic virus) and Spanish chestnuts (chestnut blight, *Endothia parasitica*) indicate the diversity and nature of the material prohibited. With such potential dangers, it is very understandable why such countries as the United States have such strict regulations governing the import of plant material. The US Department of Agriculture, for example, require an import permit for all plant material with the exception of seed, although wheat imports are prohibited from any country where flag smut (*Urocystis tritici*) is endemic unless samples pass a stringest test in a detention nursery, in which case, a permit may be granted.

There are no UK restrictions on the import of cereal seeds but export consignments might require an International Phytosanitary Certificate. In fact, the inspection and seed

Table 9.1 UK Seed Certification Standards for Loose Smut (*Ustilago nuda*)

	Minimum voluntary standard	Higher voluntary standard
Wheat and barley	(percentage by weight)	
Basic seed	Not more than 0.5% infection	Not more than 0.1% infection
Certified seed first generation	Not more than 0.5% infection	Not more than 0.2% infection
Certified seed second generation	Not more than 0.5% infection	Not more than 0.2% infection

testing of cereal grain can be a most useful exercise in determining the level of seed-borne infection and contamination.

A widely accepted method of excluding certain plant pathogens, albeit over several years, is the practice of seed certification. The statutory requirements of the UK Seed Certification Scheme are listed in 'The Cereal Seeds regulations', and 'The Seeds (Registration and Licensing) Regulations', which are regular publications of the Ministry of Agriculture, Fisheries and Food. There are both minimum and higher voluntary standards for several pathogens, and Table 9.1 lists the requirements for loose smut (*Ustilago nuda*) of wheat and barley.

The success of exclusion methods has been considerable. In the United Kingdom, the various wart disease orders (*Synchytrium endobioticum*) enforced since the original one in 1918, and the subsequent EEC regulations have, by requiring only immune cultivars of potatoes to be grown in contaminated soils plus restrictions on potato movement in and out of the area, largely eliminated this devasting disease from British agriculture.

Quarantine and allied eradication methods are credited with preventing the establishment of wart disease in West Virginia, United States, where, after a build-up of infection from 1912 to 1921, the area was quarantined with only immune cultivars being allowed. By 1974, all previously infected sites were declared free of the pathogen (Brooks *et al.*, 1974). However, the successes could easily disappear if regulations were relaxed and, in evaluating the contribution of exclusion to effective disease management, the devasting effects of late blight (*Phytophthora infestans*) in Europe, especially in Ireland in the mid-1800s, and the decimation of American chestnuts by *Endothia parasitica* and British elms by the Dutch elm disease (*Ceratocystis ulmi*) give ample warning that this management strategy is a prerequisite for the successful production of certain crops in some countries.

Eradication of the Pathogen or Reduction of the Inoculum

Although seldom achieved, the elimination of a pathogen has been the objective of many control practices. There are many ways of attempting to do this and they include the direct, physical removal of the pathogen by some cultural method or its destruction by chemical or physical means. If a disease can be recognized early during crop development, the infected plants can be physically removed, a form of eradication known as roguing. This method has been traditionally used for controlling bacterial, virus and other diseases, blackleg (*Erwinia carotovora*) of potatoes for example. Similarly, loose smut of barley (*Ustilago nuda*) and many virus diseases of cereal crops grown for seed can also be reduced by roguing. The aim of this crop hygiene technique may be to satisfy the seed crop inspectors for certification purposes although the general practice of roguing is to be recommended wherever possible if only to reduce the amount of potential inoculum.

The removal of diseased plants reached enormous proportions in the United States programme to eradicate the extremely destructive citrus canker (*Xanthomonas citri*). The disease was first observed in 1910 in the Gulf States and spread quickly through the citrus orchards. By 1934, after about twenty years of eradication, about twenty million trees had been destroyed and, although there were several sporadic outbreaks, by the 1960s there were virtually no recordings of this disease.

In the United Kingdom, similar eradication schemes since the 1960s have been successful in almost halting the spread of the beetle-dispersed Dutch elm disease (*Ceratocystis ulmi*) (Plate 9.1) although, if the outcome of the eradication schemes against this disease in the United States in the 1930s is repeated, outbreaks will still occur and result in the abandonment of the schemes.

Eradication of Alternate Hosts

The eradication of alternate hosts serves as a means for reducing the production of new physiological races (virulences) by genetic recombination in such pathogens as the rust fungi where the sexual stage occurs on alternate hosts. A good example is the heteroecious pathogen *Puccinia graminis tritici* which causes black stem rust on wheat. This pathogen alternates between the wheat plant and the common barberry bush (*Berberis vulgaris*). Many barberry eradication schemes have been introduced, the first in France in 1660; the most publicized was the continuing scheme started in 1918 in the United States in which, by 1942, around 300 million bushes had been destroyed. The scheme succeeded to the extent of reducing the early spring inoculum and delaying the epidemics but never achieved complete success due, in part, to the incomplete barberry eradication and also to the long distance dispersal of uredospores from the southern states where they over-winter.

Eradication of Alternative Hosts

The eradication of alternative hosts will also reduce the potential number of infected plants. Alternative hosts provide the pathogen with a means of overwintering when its economic host is an annual. Cruciferous weeds such as charlock and shepherd's purse are alternative hosts for the club root pathogen (*Plasmodiophora brassicae*) and many weed grasses, *Agropyron repens* for example, can carry the take-all pathogen through the intercrop period in which a non-susceptible crop is being grown. Many virus diseases of cereals, such as barley yellow dwarf, have weed grasses as alternative host species and chickweed (*Stellaria media*) has been shown to be a symptomless carrier of cucumber mosaic virus (Tomlinson *et al.*, 1970). Volunteer cereal plants also carry over inoculum of the powdery mildews, rusts and *Septoria* spp. during the intercrop period and, although this period may only bridge the gap between harvest and the subsequent autumn-sown crop, great care should be taken to remove and destroy them.

Cultural Methods

Cultural methods of eradication include many mundane tasks which, superficially, appear to be only justifiable on aesthetic grounds but, epidemiologically, they attempt to prevent disease by the elimination of a major inoculum source. Such is the case with the hygienic practice of destroying the leaves of apples and pears on the orchard floor during the winter. If this is carried out efficiently, by either burning or chemical means, the sexual stage of the apple scab fungus (*Venturia inaequalis*) will be destroyed. In similar manner, the pruning of apple trees to cut out scab-infected bark, twigs infected with canker (*Nectria galligena*), pear branches infected with the fire-blight bacterium (*Erwinia amylovora*) or

Plate 9.1 Dutch elm disease. Contrast almost dead tree in background with healthy tree in foreground. (Photo: E. Wintle)

plum and cherry branches with 'hold-over' cankers of *Pseudomonas mors-prunorum*, are highly recommended activities for the winter in commercial orchards.

Burning-off Potato Haulms
The burning-off of potato haulms prior to harvest can be effected by many chemicals, traditionally based on sulphuric acid but more recently on desiccants such as metoxuron. This treatment is a most effective method of eradication of the late blight pathogen (*Phytophthora infestans*) from within the foliage and, at the same time, prevents the spread of sporangia from the leaves into the soil where they could infect the tubers and perpetuate this 'seed-borne' disease.

Eradicating Pathogens by Rotation
Cultural methods of control also include the eradication of the pathogen by rotation. The length of the rotation varies with the longevity of the pathogen and may be as short as one year, as in the case of the take-all pathogen of cereals (*Gaeumannomyces graminis*) which, being a poor soil competitor will disappear with the decay of the host debris, or as much as seven years in the case of club root (*Plasmodiophora brassicae*) whose long-lived resting spores pose a considerable rotational problem if soils are contaminated.

Destroying Plant Debris
Many cultural practices are aimed at the destruction of plant debris which may be harbouring pathogens. Good ploughing techniques not only contribute to improved soil structure, but if carried out sufficiently early in the autumn also incorporate infected plant material into the soil before temperatures drop too low to halt the otherwise rapid decomposition of the organic matter. Many cereal pathogens, *Septoria nodorum*, *Rhynchosporium secale* and *Erysiphe graminis* can be starved of nutrients in this way as can debris-borne pathogens of many other crops, stem canker (*Leptosphaeria maculans*) of oilseed rape, for example. With a similar objective, the burning of cereal stubble will reduce the overwintering cleistothecial stage of *Erysiphe graminis* as well as many other potentially destructive pathogens. Crop hygiene of this nature will, inevitably, be less than totally effective and the law of 'diminishing returns' is always in operation in that it is probably not cost-effective to make the additional effort to destroy every minute plant fragment. However, with polycyclic pathogens which can multiply rapidly, severe epidemics can still develop from very small amounts of initial inoculum if conditions are conducive to disease development.

Applying Chemical Seed Treatments
The application of many chemical seed treatments also qualify as eradicants and there can be no better examples than the traditional use of organomercurial dusts for the control of such diseases as bunt of wheat (*Tilletia caries*) whose spores contaminate the grain. With the banning of mercury compounds by many countries, the use of carboxin and guazatine formulations have met with much success.

The eradication of pathogens present in soil is, if anything, increasing with the availability of more efficient chemicals and better technology for their application. In this context, the use of many much used soil sterilants including formaldehyde, methamsodium, methyl-isothiocyanate and dazomet must be mentioned as should the more expensive and labour-intensive use of steam sterilization.

Protection of the Host

Crop plants may be protected against invasion by pathogenic organisms either by possessing efficient resistance mechanisms or by the application of chemicals. Disease resistance is dealt with elsewhere in this book (Chapter 6) and only chemical application will be considered here. Although there is much opposition from the environment lobby to the use of chemicals in agriculture, there can be no doubt that the world food situation has benefited from the availability of chemicals to control many pests and plant pathogens. Success stories are plentiful although the chemical control of viruses still presents problems to the agrochemical industry.

Chemical Control

The term 'chemical control' requires qualification as it implies the elimination of a population of harmful pests or pathogens. In practice, the aim of most management strategies involving chemical applications is simply to reduce disease levels to the extent that the grower both increases his crop yield and more than recoups the extra cost of the chemical treatment.

Chemicals were used to control plant diseases long before the nature of their causal organisms had been elucidated. The early writings of Homer clearly described the use of the 'all-curing' sulphur in Greek agriculture but, because of ignorance of the nature of the maladies of the plants being treated, the foundations of the agrochemical industry were not layed down until the nineteenth century when Prévost used copper sulphate to control the seed-borne bunt pathogen (*Tilletia caries*). The momentum increased with the introduction of lime-sulphur by H. Marshall Ward in 1880 and the development of the copper-based Bordeaux mixture by Millardet in France in 1885. The fascinating story of the early days of fungicide research is most interestingly recounted by E. C. Large in *The Advance of the Fungi* (1940).

Grouping Chemicals

Chemicals can be grouped according to the objective of their use:

(1) the protection of the healthy host (**prophylaxis**),
(2) the cure or therapy of the diseased host (**chemotherapy**),
(3) the destruction of spores or pathogen propagules on the host surface, e.g. seeds (**disinfestation**).

By this categorization, we can recognize three types of chemical, the **protectant**, the **eradicant** and the **disinfectant**. The modern systemic is not an additional category as will be discussed later as, in reality, its effect is one of eradication albeit of 'young' infections.

By definition, the protectant is applied to the healthy host plant before the arrival of the inoculum. Hopefully, it will supplement the host's own defences by forming a superficial chemical barrier. To achieve this, the chemical must be present in high enough concentration when the inoculum arrives. This implies that it must be exceptionally well distributed over the exposed plant surface, it is stable and possesses properties of high initial retention and tenacity of residue. Weathering will be its main limiting factor and, more than likely, a programme of sequential applications will be required to maintain a high level of protection unless disease forecasting can precisely predict the time of inoculum arrival and hence the time to spray.

Genuine eradicants are rare and, except for systemic chemicals, their use is very limited because it is usually too late for chemical control when infection is firmly established and colonization well in progress.

Chemical disinfectants are most usually applied to the seed or to the environment, mainly the soil immediately surrounding the host plant.

Of the chemicals used in disease management strategies, fungicides far outnumber bactericides, insecticides, nematicides and viricides. Fungicides can be used in many forms but it is convenient to group them into five major formulation types;

(a) suspension concentrates,
(b) wettable powders,
(c) dispersable powders,
(d) emulsifiable concentrates, and
(e) grains/granules.

No matter in what form the chemical is used its effect on fungi will be either:

(1) *fungicidal*, where the fungus is killed by the chemical,
(2) *fungistatic*, where fungal growth is inhibited, or
(3) *genestatic*, where fungal sporulation is reduced.

The Application of Fungicides

Ideally, chemists and pathologists would like a chemical that could be applied as a seed-dressing or be rapidly absorbed by plant roots. In this way, the growing plant might be protected throughout its susceptible period and, if the chemical were sufficiently broad-spectrum, this protection could act against many pathogens. This ideal may never be attained although the advent of xylem translocated systemic fungicides has improved the situation. A downwardly translocated (phloem transport) chemical would very much increase the efficiency of internal distribution of foliar application.

Chemicals can be applied in three main ways, as dusts, sprays or fumigants. The main objective of dusting or spraying a crop is to produce a chemical barrier which protects the susceptible tissue. The attainment of good distribution and coverage is essential for conventional protectant compounds although this is less important with systemic chemicals. Problems of distribution and coverage do not apply when using fumigants but, as a rule, these chemicals are not persistent.

Dusting

Potentially, there are several advantages of dusting over spraying. Not least of these is the fact that water is not a requirement but the lower weight and greater manoeuvrability of dusting machines also contribute to their merits. Disadvantages also exist, the requirement for very fine particles in the formulated product often causing 'balling' due to static electroforces, and also to 'caking' if the dust absorbs too much moisture. The nature of the leaf surface significantly affects retention of the applied chemical as is instanced by the greater deposition on abaxial leaf surfaces, presumably because of the prominent vein structure. The optimum time to apply a dust is when the leaves are still moist after rain or dew, but dusting becomes very inefficient if not carried out on the calmest days.

Spraying

This is by far the most widely practised method of applying chemicals for disease control with water being the usual carrier although light oils are used occasionally. There are several kinds of spray machinery, each characterized by the spray volume applied:

High volume (HV) > 700 l/ha
Low volume (LV) $50-200$ l/ha
Ultra-low volume (ULV) < 50 l/ha (usually < 5 l/ha)

Plate 9.2   An Allman Unibilt 625 special tractor mounted sprayer. (Photo: E. Allman & Co Ltd)

The use of a **high-volume sprayer** requires a plentiful water supply as its large tank will probably have a 500 l capacity (Plate 9.2). The whole machine will normally be mounted on the rear of a tractor, the spray chemical being delivered via long booms along which are placed many nozzles. Traditionally, the high-volume sprayer would be used to deliver the chemical until the 'runoff' point is reached although this is not absolutely necessary if the distribution is uniform.

Low-volume sprayers were introduced to reduce the water requirement and, with only about 200 l/ha being the more normally applied volume, the weight of the machinery is much less than its high-volume counterpart. The low-volume sprayer uses a pressure pump to produce an atomization effect, the resultant deposit being, hopefully, a uniform coverage of fine droplets.

The progress of tractor and spraying machinery through a cereal field on several occasions to apply protectant chemicals causes damage which discourages such application. The advent of systemics and the **tramline** system of farming, in which the tractor follows the same tracks for each 'in-field' operation, has revolutionized management strategies in cereal crops with 60–70% of wheat and barley crops in the United Kingdom now being regularly sprayed.

In general, small droplets are highly desirable if uniform and complete coverage is to be obtained. The disadvantage of small droplet application is spray drift, accentuated if aeroplanes or helicopters are used. However, low-volume spraying is not considered to be as efficient in terms of fungicide usage as high-volume spraying, and Fry (1982) quotes the example of applying fungicide in 450 l/ha which suppressed late blight of potatoes to 10–20% of the level resulting from applying the same volume in 22 l/ha.

The use of slowly rotating discs or cups to produce a fine, atomized spray application has, with the adjustment of low rates, enabled a greater control of droplet size. This method has introduced a new term into spray technology, **controlled droplet application** (CDA), and it is hoped that this method will ultimately lead to a more efficient usage of

Plate 9.3 Controlled droplet application
   (i) CDA Boom sprayer fitted with cable driven Micromax multi-speed heads spraying for aphids
        in barley and applying 100 micron droplets.
   (ii) The Micron Micromax spinning disc head.
(Photos: Micron Sprayers Ltd)

chemicals (UK MAFF Advisory Leaflet 792). The machinery for this type of **ultra low-volume spraying** can either be of the hand-held type or tractor-mounted (Plate 9.3). All aspects of these developments are the subject of two excellent reviews which are discussed in considerable detail by Mathews (1979).

The concept of using electrical charges to improve the impaction of fine particles onto foliage is, at present, very much under review with much evidence to suggest considerable potential in this technique.

Fumigation

The early fumigants were all used in enclosed spaces like glasshouses and the traditional method of painting sulphur on to the hot water pipes is a good example. More sophisticated apparatus is now available, the controlled electric heater for volatile products, the pyro-technic smoke generator to disperse the volatile toxicant and various forms of pressurized 'aerosol' applicators.

Fumigants are also used for purposes of soil-sterilization, often of batches of soil brought into special sheds and also outside and applied before the crop is sown. The formulation may be as granules or emulsions and often the soil is watered after application to seal the surface against vapour loss. Today, it is quite common for soil to be sterilized on quite an extensive scale utilizing large drag sheets, usually of polythene or similar compound. Quite obviously, great care must be taken to protect the operator in fumigation treatments, the soil will probably require a period of perhaps up to six weeks to dissipate the chemical and, with considerable expense involved, only soil to be sown to high cash crops such as strawberries and tomatoes justify this treatment.

Formulation Additives

Various substances can be added to dusts and sprays to improve their initial retention and tenacity of residue. Such compounds or stickers certainly do enhance retention although this does not always lead to an increase in efficacy (Ogawa, Gilpatrick and Chiarappa, 1977). Many factors affect the sticking quality of fungicides, fine particles being more tenacious than large particles, but additives such as oils, natural and synthetic adhesives and pastes have all been used in this context.

Similarly, adjuvants are often added to the toxicant to overcome problems of dis-tribution on the leaf surface or to overcome any hydrophobic qualities of certain waxy leaves. Spreaders or surfactants (wetters) are used to this effect and may be simple forms of detergents.

The Timing of Application

The application of chemicals is always an expensive operation and every effort should be made to ensure maximum efficiency. With seed treatments, the timing of application is not particularly critical although the nearer to sowing the operation is carried out the less chance there will be for loss of the chemical by rubbing or by degradation. With foliar applied fungicides, efficiency will depend very much on the timing of application. The objective with protectant fungicides is to make the applications as near to the arrival of the inoculum as possible. The classical experiments with the timing of application of copper fungicides to control potato blight are good examples of this principle (Evans, Couzens and Griffiths, 1965). Frequently applications may also be necessary to counteract the loss of chemical due to weathering but disease forecasting, as discussed previously (Chapter 8) is essential to ensure the optimum timing for the first application.

In many cases, the decision when to spray will involve a consideration of epidemiology, geographical location and possible yield loss. With barley mildew (*Erysiphe graminis hordei*) the main aim is to protect the topmost leaves and ears with a systemic fungicide. Commonly, an application of triadimefon, tridemorph or ethirimol is made in the spring with a second application if forecasts indicate the likely build-up of disease. Early infections of the wheat glume blotch disease (*Septoria nodorum*) are often ignored, the optimum spray timing being between GS8–10 (decimal key 37–43) which is prior to the 'in boot' stage and will give protection to the emerging head.

Fungicide Resistance

The advent of systemic chemicals, although much welcomed for efficient disease management treatments, also introduced a phenomenon which brought with it sinister undertones. The occurrence of fungicide resistance in plant pathogens had, hitherto, been an extreme rarity, but the appearance of strains of *Sphaerotheca fuliginea*, the causal organism of powdery mildew of cucurbits which were resistant to dimethirimol, and of strains of *Botrytis cinerea*, the grey mould pathogen resistant to benomyl, certainly alerted chemists and pathologists to this alarming development.

The hazard of fungicide resistance varies with the type of chemical in use (Dekker, 1981). The 'conventional fungicides' such as copper compounds and dithiocarbamates are not 'single-site-directed' and do not exert selection pressure for the emergence of mutant isolates. Such 'multisite' fungicides appear to offer no real problems, although resistance caused by decreased uptake or increased detoxification through metabolism is possible. The risk of failure of disease control in these multisite inhibitors is thus relatively low, although resistance to organic mercury has been reported in the oat leaf pathogen, *Pyrenophora avenae* (Noble *et al.*, 1966).

Mutant strains of *Botrytis cinerea* resistant to dicarboximides have been regularly found but often with no reduced disease control (Beever and Byrde, 1982). The key factor in the build-up of resistant strains appears to be their reduced fitness. In the United Kingdom, surveys have indicated that barley powdery mildew isolates (*Erysiphe graminis hordei*) varied considerably in their 'sensitivity' to ethirimol, but the resistant isolates were less fit and declined in frequency in areas where application of this chemical ceased.

Resistance to benzimidazoles or MBC-generating compounds, although reported for certain pathogens, was thought to be unlikely in the eyespot pathogen of cereals (*Pseudocercosporella herpotrichoides*) because of epidemiological considerations such as limited dispersal. However, trials in the United Kingdom in 1982 on sites in which severe lodging had occurred in 1981 on MBC-treated crops of winter wheat showed that eyespot was not controlled by one or more sprays of an MBC fungicide. By 1984, resistant strains were quite widespread and management strategies of fungicide diversification had to be adopted.

Much more information is required about resistance mechanisms in field-resistant isolates. It is known that cross-resistance patterns occur between chemicals having the same mode of action. The morpholine fungicides tridemorph and fenpropimorph inhibit sterol biosynthesis but at a different step to that affected by triadimenol and related fungicides. As yet, no cross-resistance between morpholine and triazole fungicides or ethirimol has been found.

Fungicide resistance poses a risk to disease management and, to improve the scientific base for management decisions, much research is now being put into strategies aimed at prolonging the useful life of fungicides. An excellent review of fungicide resistance problems by Staub and Sozzi (1984) discusses such strategies. One approach would be to use fungicide mixtures where each constituent had a different mode of action. Another approach is to alternate the use of different fungicides either on a spray-by-spray basis or on a field basis. A mixture of metalaxyl and mancozeb was far superior to metalaxyl alone in the inhibition of the frequency of occurrence of resistant strains of *Phytophthora infestans* clearly indicating that mixtures should delay the build-up of resistance, a view forwarded by Kable & Jeffery (1980) using mathematical models. Only time and further experimentation will reveal the best application strategies and this will undoubtedly include a diversification of the fungicides in time and/or in space.

Adjustment of the Environment

Disease is the result of the interaction of the host plant, the pathogen and the environment and, whilst nothing can be done to manipulate prevailing weather conditions, much can be done to effect changes in the microclimate which will adversely affect disease development. Even the conditions for seed germination, emergence and growth should be carefully monitored to encourage the rapid and vigorous early growth necessary to give maximum likelihood of attaining healthy maturity. Planting in properly cultivated soil of suitable texture, seed-bed compaction, pH, moisture content, fertility and temperature is essential. Depth of sowing, seeding rate and over-watering can have very deleterious effects on seedling vigour and susceptibility to attack and the all-too-familiar damping-off disease (*Pythium de Baryanum*) is the penalty for producing these disease conducive conditions. Sowing date in autumn-sown cereals also has important implications in terms of disease although there is always a compromise to be reached between early sowing to produce a well-established crop by the onset of winter and the greater chance of being infected by disease propagules, spores of powdery mildew or rust, for example, which have survived since the summer harvest.

Adjustment of the environment can be attempted both in the field and in the glasshouse. Liming is a common practice in soils where club root may be a problem as the pathogen, *Plasmodiophora brassicae*, is much less likely to infect in alkaline soils. Similarly, manipulation of irrigation water has also been shown to adversely affect disease. In Israel, overhead sprinkling of tomatoes reduces powdery mildew (*Leveillula taurica*), but increases the chance of infection by the bacterium *Xanthomonas vesicatoria*. Manipulation of glasshouse humidity by ventilation is also known to be effective in reducing tomato leaf mould (*Fulvia fulva*) and grey mould of tomatoes caused by *Botrytis cinerea*.

Physical and Biological Management Strategies

Although very much in the minority, physical and biological management strategies can be most important, often offering the only possibility of reducing disease potential. Heat has been used in a variety of ways. Loose smut of wheat and barley (*Ustilago nuda*) is systemic in the seeds and, prior to the advent of systemic chemicals, infected seeds would be submerged in hot water at a temperature high enough to kill the pathogen without reducing the viability of the seed. A common practice was to presoak the seed for 4–6 hours followed by a further immersion at 54°C for 10 minutes. The mechanism for destroying the fungus is not fully understood but it is thought that antibiotic substances produced during anaerobic respiration may be responsible.

Heat is also used to eliminate some viruses from planting material. By growing plants at about 34–36°C for several weeks, many viruses of vegetatively propagated horticultural species can be eliminated. Hollins (1965) reports that valuable cultivars of raspberry, strawberry and grapevine have been revitalized in this way. Heat treatment is also used in conjunction with apical meristem and tip culture. Apical meristems of infected plants usually contain little or no virus. If the plant is pre-heated before use, more tissue will be freed of virus allowing a larger portion of shoot tip to be taken giving it a greater chance of establishing in culture and eventually growing into a mature plant. This technique has become standard with plants such as carnations, chrysanthemums and potatoes and is described in detail by Gibbs and Harrison (1976).

Steam sterilization of soil is carried out in many glasshouses to reduce contamination by soil-borne pathogens such as *Pythium*, *Fusarium* and *Verticillium*. The success of the treatment depends upon the soil being uniformly heated to about 100°C for about 20 minutes. Raising the ambient temperature to 24°C within a glasshouse is also carried out to reduce attack from *Verticillium* spp.

Much has been written about the potential of biological control of plant diseases (Baker and Cooke, 1974). However, only two examples have really established themselves as commercial practices. The crown gall bacterium, *Agrobacterium tumefaciens* has a wide host range, being a particular problem on cuttings and seedlings of peaches, apples and other fruit trees. Pathogenic strains of *A. tumefaciens* can be overcome by a bacteriocin-producing strain of *A. radiobacter*. The bacteriocin from strain 84 is called agrocin, and seedlings or cuttings dipped in a cell suspension of this strain are protected by this compound (Kerr and Htay, 1974).

The classical example of biological control is the application of a spore suspension of the fungus *Peniophora gigantea* to the stumps of recently felled trees to prevent infection by the damaging pathogen *Heterobasidium annosum (Fomes annosus)*. Rishbeth (1950) elegantly demonstrated that prior colonization of the stumps by the saprophytic *P. gigantea* was highly effective in controlling this disease.

Major Groups of Fungicides

Inorganic Fungicides

Copper

Copper compounds were first used in 1807 by Prévost for the seed treatment of wheat and are still very much used today despite the steady decline in the use of the classic Bordeaux mixture which established copper as a foliar fungicide in 1885. It was discovered by Millardet at Médoc in the Gironde, France, and henceforth was given the name **Bordeaux mixture**. It is a mixture of copper sulphate and hydrated lime and was first used against downy mildew of vines (*Plasmopora viticola*). Copper fungicides are protectants, exposed to weathering and normally require a programme of sprays. They are often corrosive and somewhat phytotoxic, certain apple cultivars for example become russetted, the **copper shy** cultivars. They can also reduce photosynthesis and translocation by their colour and consistency. Bordeaux mixture has now been largely replaced by what are termed 'fixed copper' fungicides, mainly the basic sulphates and chlorides and the oxides. Copper has been mixed with sodium carbonate (*Burgundy mixture*), ammonium hydroxide (*Azurin*), ammonium carbonate (*Chestnut compound*) and as several dust formulations.

Copper fungicides are widely used as sprays against downy mildews, especially potato late blight, and a variety of other leaf-spotting diseases.

Sulphur

Sulphur fungicides are based on either elemental sulphur or various polysulphides. Elemental sulphur is used as a dust or may be formulated as a wettable powder; the fungitoxicity of elemental sulphur is dependent upon particle size, small particles being most effective. Sulphur is phytotoxic and, though lime-sulphur largely replaced elemental sulphur, even this somewhat complex mix of calcium polysulphides and calcium thiosulphate can be phytotoxic on certain crop cultivars—the **sulphur-shy** cultivars. Although sulphur-based fungicides have now been replaced for many purposes by less toxic organic fungicides, they are very efficient in the control of apple scab and the powdery mildews, especially powdery mildew of grape (*Uncinula necator*).

Mercury

The two chlorides of mercury—mercurous chloride (Hg_2Cl_2) and mercuric chloride ($HgCl_2$)—have found a number of applications in the control of fungal disease. However, all formulations are extremely toxic to man, animals and birds and there is much public opposition to their use. In many countries, New Zealand for example, mercury-based fungicides have been banned by legislation. Organomercury fungicides have the general formula R—H_2—X, where R may be either an aryl or alkyl radicle and X an anion such as phosphate or chloride. They have been mainly used as seed dressings, especially for the control of bunt (*Tilletia caries*) and the covered smuts, and are particularly effective against seed contaminants such as *Fusarium* and *Helminthosporium*. As phenyl mercury acetate or chloride they are also efficient as eradicants of recent apple scab infections (*Venturia inaequalis*), but they must be applied within 72 hours of infection to be effective.

Organic Fungicides

In the search for less toxic compounds than the copper and sulphur fungicides, the **dithiocarbamates** patented by the Du Pont Company in 1934 have been very successful in meeting this need.

The first of these compounds, tetramethylthiuram disulphate, now known as **thiram**, quickly became established both as a foliar protectant of turf grasses and tulips and as a cereal seed dressing.

Thiram

The iron and zinc salts of dithiocarbamic acid, **ferbam** and **ziram** were reasonably successful but have largely been replaced by **zineb** and **maneb**, the two latter fungicides being the zinc and manganese salts which have been most important in the chemical control of vegetable diseases, particularly late blight of potatoes (*Phytophthora infestans*) and early blight of tomatoes (*Alternaria solani*).

The sodium salt, **nabam**, was first marketed under the trade name Dithane as a protectant foliar spray and, if mixed with zinc sulphate, when it is called zineb tank mix, it gives good control of tomato leaf mould (*Fulvia fulva*). Nowadays, all the metallic bisdithiocarbamates are marketed as various types of Dithane, the most recent being a complex of zinc and maneb containing 20% manganese and 2.5% zinc, now called **mancozeb**.

One member of the dithiocarbamate group has found an important application as a soil fungicide. This is **metham sodium** (sodium N-methyl dithiocarbamate).

Metham sodium

In the soil, it produces methyl isothiocyanate which is considered to be the active ingredient; but this can also be produced by another dithiocarbamate,

$$
\begin{array}{c}
\quad\quad S \\
H_2C \diagup \quad \diagdown C{=}S \\
\mid \quad\quad\quad \mid \\
CH_3N \diagdown \quad \diagup NCH_3 \\
\quad CH_2
\end{array}
\qquad \textit{Dazomet}
$$

Other Organic Fungicides

Captan, a phthalimide, has proved a particularly effective fungicide for the control of apple scab. It inhibits thiol-containing enzymes and may also react with sulphydryl groups.

$$
\begin{array}{c}
\quad CH_2 \\
HC \diagup \quad \diagdown CH{-}CO \\
\mid \quad\quad\quad \mid \quad\quad\quad \diagup NSCCl_3 \\
HC \diagdown \quad CH{-}CO \\
\quad CH_2
\end{array}
$$

(N–[trichloromethylthio]–
cyclohex-4-ene-1, 2-   *Captan*
dicarboximide)

Captafol is a closely related compound and is highly effective against many diseases including blight of potatoes and coffee berry disease (*Colletotrichum coffeanum*). However it is a skin irritant, affecting a small proportion of sensitive individuals. It is a very persistent, non-systemic protectant and is sometimes used in mixtures with systemic MBC-generating compounds for the control of *Septoria nodorum* on wheat.

Dinocap is a dinithrophenol, a group of compounds recognized to have useful properties as herbicides and insecticides as well as fungicides. It has found a popular use as a protectant against powdery mildews, in particular powdery mildew of apples (*Podosphaera leucotricha*).

Dinitro-orthocresol (DNOC) is a closely related compound and has been used for spraying orchard floors as part of a programme to eradicate the apple scab fungus from fallen leaves.

Miscellaneous Fungicides

Organotin compounds are fairly broad-spectrum in activity. Two compounds have found reasonable markets, triphenyl tin acetate (fentin acetate) and triphenyl tin hydroxide (fentin hydroxide). They have found favour in the control of potato blight having an added advantage over competitors in this area by giving protection of the tubers as well as the foliage.

Systemic Fungicides

The fungicide manufacturers reaped a just reward in the 1960s and 1970s when after years of expensive and painstaking research they developed chemicals that could penetrate the plant cuticle and then be translocated within the plant. Some chemicals may be absorbed but not translocated, in which case they can only confer fungitoxicity at the sites of application. These chemicals are termed **topical therapeutants** or eradicant fungicides but will not be dealt with in this section. A good general review of systemic fungicides was written by Marsh (1977).

The systemic fungicide has distinct advantages over the eradicant and traditional protectant in that it is internally distributed and hence not been subject to weathering. In addition, application need not be undertaken until after infection has occurred and, in consequence, eliminates the costly wastage inherent in incorrectly timed insurance spray programmes where the onset of disease cannot be predicted accurately.

Penetration of the cuticle is the first obstacle to the uptake of systemic fungicides and this appears to be successfully achieved if the chemical is water-soluble. The chemical is then able to pass via the intercellular spaces to the vessels and upwards through the plant. Application of a systemic fungicide to the root has the best potential to ensure complete distribution through plant tissues. With downwardly transported (phloem) chemicals being almost non-existent, it follows that application to the shoot is likely to result in incomplete distribution. Shoot application, however, is relatively easy, can be repeated and allows reasonable dosages. Root application in the form of granules, soil drenches or seed treatments does not have these advantages and the relatively small amounts that can be applied as seed treatments will inevitably be rapidly diluted as the plant grows, almost ensuring that a second, foliar application will be required if the mature plant is to be protected. To illustrate this point, in the United Kingdom spring barley is often seed-treated with ethirimol which gives excellent protection against powdery mildew (*Erysiphe graminis hordei*) for much of the season. However, current recommendations are to spray the foliage at GS39 (decimal key) if 3.5% of the area of the oldest green leaf is damaged by this disease.

Systemic fungicides are extensively used in the control of cereal diseases (Jones and Clifford, 1983) both as foliar sprays, such as triadimefon, carbendazim or tridemorph, and as seed-dressings, such as carboxin for the control of loose smut (*Ustilago nuda*). Benomyl is a very broad-spectrum systemic, used as a spray for the control of eyespot of wheat (*Pseudocercosporella herpotrichoides*) or apple scab (*Venturia inaequalis*), and also as a dip for the control of storage rots of citrus fruits (*Penicillium italicum* and *P. digitatum*). In Japan, granules of the fungicide **kitazin**, an organophosphorus compound, are applied to the irrigation water to control rice blast (*Pyricularia oryzeae*), and there are many references in the literature to the use of benomyl and thiabendazole for the control of soil-borne pathogens of other crops. Lastly, a recent development has been the application of systemic chemicals in the water supply to glasshouses using the nutrient film techniques; for example, etradiazole is used to control *Pythium* root rot in tomatoes.

Systemic Fungicides in General Use
The following comprise the main groups of fungicides in current use.

Benzimidazoles   These are normally designated 'broad-spectrum' but, in fact, *Basidio-mycetes* and *Phycomycetes (Mastigomycotina)* are generally quite insensitive. Many are known as 'MBC generators' as they owe their activity to methyl benzimadazol-1-yl carbamate (MBC) to which compound they decompose within the plant or in water *in vitro* although **carbendazim** is the MBC compound itself. This group includes **benomyl, carbendazim, fuberidazole, thiophanate-methyl, thiobendazole (TBZ)**.

Ergosterol Biosynthesis Inhibitors   Here the activity is directed at the disruption of synthesis and function of cell membranes by the inhibition of ergosterol biosynthesis. The group includes **imazalil, triforine, prochloraz, propiconazole, triadimefon, triadimenol**. The last four are members of the triazole group and represent some of the more successful systemics used against cereal diseases.

Morpholines   Again, these are ergosterol inhibitors but their mode of action is different. The group includes **fenpropimorph, tridemorph, dodemorph** (roses).

Hydroxypyrimidines   These are highly selective against powdery mildews and include two of the earliest systemic fungicides. The group includes **dimethimirol** (for cucurbits), **ethirimol** (for cereals), and **bupirimate** (for roses and apples).

Organic Phosphates   This group includes **ditalimfos** (used against barley powdery mildew), **kitazin** (used in the Far East for the control of rice blast), and **tolchlofos methyl** (used against *Rhizoctonia* on potatoes).

Carboxamides   These are closely related to the morpholines, and are more often referred to as **oxathin** derivatives. The group includes **carboxin** (used extensively against loose smut), and **oxycarboxin** (the sulphone derivative used against cereal rusts).

Dicarboximides   This group includes **iprodione** and **vinclozolin** used in the control of diseases of oil-seed rape in particular.

Guanidines   This is a small group which includes **guazatine** used in seed treatment against bunt, etc.

Miscellaneous

This includes **chloroneb**, which has limited systemic action, but control of *Rhizoctonia*, *Pythium*, *Sclerotium*, *Typhula*; **prothiocarb**, which is used as a soil drench or as an in-furrow spray near the roots of crops against *Pythium*, *Phytophthora* and *Pseudoperonospora*; and **metalaxyl**, which is a member of the acylalanine group. It is selectively toxic to *Oomycetes*, and is widely used against potato late blight, though resistant strains are a problem.

Footnote

As crop yields approach their theoretical maxima, as calculated by plant physiologists and agronomists, only the reduction of losses due to plant diseases offers the possibility of increasing the actual yields obtained. In some situations, especially in developing countries there will be much scope for reducing disease by utilizing one or several of the management strategies outlined in this chapter. In the more sophisticated agricultural systems of the more developed countries, especially in Europe and the United States, other pressures might limit the choice of management technique. With crops in which large surpluses are now being achieved, there will inevitably be political pressure to reduce such over-production. Commodity price control might be one approach and crop quotas another but, whatever method is implemented, a reduction in the cost of inputs to the crops appears to be unavoidable. Fertilizers and pesticides would seem to be choice candidates for input reduction, both also suffering from the additional pressures exerted by the growing environmentalist lobby.

The above situation does not necessarily spell 'doom and gloom' for the agrochemical industry but it might well result in a change of emphasis in terms of disease control strategies. A more integrated approach is always to be recommended with all plant disease problems and the pitfalls of concentrating on single, highly specific control measures have been amply demonstrated by the rapid breakdown of race-specific resistances and the insensitivity of certain pathogens to some site-specific, systemic fungicides. The likely and more sensible approach will be to adopt strict cultural control measures in which crop hygiene and rotation are of paramount importance. Disease resistance will certainly be increasingly encouraged although it is too early to evaluate the part that might be played by the latest developments in biotechnology in this respect. It can only be assumed that the intensive use of agrochemicals will be limited by choice or by financial penalty but, if used in a planned, rational manner, integrated with host diversification schemes to conserve useful resistance genes, agrochemicals will still enable crop diseases to be managed efficiently, economically and with the minimum social and ecological disturbance.

PART II

Compendium of Diseases

The selection of diseases for any textbook in plant pathology will inevitably be subjective, by definition be incomplete and suffer from reviewer's criticisms of having a parochial bias. The selection of diseases for an introductory text will not escape these condemnations and may even compound the deficiencies by the additional limitations of space.

It is hoped that the diseases included in this compendium will, at the very least, merit inclusion as being of worldwide distribution and importance and, at the same time, fulfil the requirements of a taught course in covering the diversity of pathogen groups, disease symptoms and mycological idiosyncrasies. No attempt has been made to give comprehensive coverage to plant pathogens and each disease covered could well have been described in much more detail. It can only be hoped that twenty-five years teaching experience has guided me to include the diseases which best illustrate the principles of taxonomy, symptomatology, epidemiology and approaches to control.

Each disease is described from the point of view of field identification, occurrence and economic importance, life-cycle and pathogenicity and, where appropriate, details are given that might be helpful in laboratory investigations. Control measures are included with the emphasis on integrated strategies as, although specific treatments may be described, current recommendations are often of an ephemeral nature and may quickly be outdated.

The fungal diseases described in this compendium have been arranged in an order which reflects their taxonomic position, starting with the Myxomyota and then proceeding through the Eumycota from Mastigomycotina to Deuteromycotina (see Table 3.1). This section is then followed by diseases caused by bacteria and by viruses, both arranged in alphabetical order.

List of Diseases

Diseases Caused by Fungi

Club root of crucifers: *Plasmodiophora brassicae*
Powdery scab of potatoes: *Spongospora subterranea*
Wart disease of potatoes: *Synchytrium endobioticum*

Watery wound rot of potatoes: *Pythium ultimum*
Late blight of potatoes and tomatoes: *Phytophthora infestans*
Downy mildew of tobacco: *Peronospora tabacina*
Lettuce downy mildew: *Bremia lactucae*
Peach leaf curl: *Taphrina deformans*
Cereal powdery mildew: *Erysiphe graminis*

Other powdery mildews

Apple and pear scab: *Venturia inaequalis* (*V. pirina*)
Ergot of cereals and grasses: *Claviceps purpurea*
Take-all or whiteheads of cereals and grasses: *Gaeumannomyces graminis*
Glume blotch of wheat and barley: *Leptosphaeria nodorum*
Net blotch of barley: *Pyrenophora teres*
Southern leaf blight of maize: *Cochliobolus heterostrophus*
Cottony rot of vegetables: *Sclerotinia sclerotiorum*
Common bunt or stinking smut of wheat: *Tilletia caries*
Loose smut of wheat and barley: *Ustilago nuda*
Common smut of maize: *Ustilago maydis*
Black stem rust of cereals: *Puccinia graminis*

Other rust fungi

Grey mould of lettuce: *Botrytis cinerea*
Seedling blight, foot and root rot and head blight of cereals: *Fusarium culmorum*
Verticillium wilt of tomatoes: *Verticillium albo-atrum and V. dahliae*
Eyespot of wheat and barley: *Pseudocercosporella herpotrichoides*
Leaf blotch or scald of barley: *Rhynchosporium secalis*
Tomato leaf mould: *Fulvia fulva*
Leaf-spot of celery: *Septoria apiicola*

Diseases Caused by Bacteria

Crown gall: *Agrobacterium tumefaciens*
Fire-blight of apples and pears: *Erwinia amylovora*
Blackleg of potatoes: *Erwinia carotovora* subsp. *atroseptica*
Bacterial wilt of maize: *Erwinia stewartii*
Bacterial canker of tomato: *Corynebacterium michiganense*
Bacterial ring rot of potatoes: *Corynebacterium sepedonicum*
Yellow slime or 'tundu' disease of wheat: *Corynebacterium tritici*
Halo blight of oats: *Pseudomonas coronofaciens*
Halo blight of dwarf and runner beans: *Pseudomonas medicaginis* subsp. *phaseolicola*
Bacterial canker or shothole of plums and cherries: *Pseudomonas mors-prunorum*
Black rot of crucifers: *Xanthomonas campestris*

Diseases Caused by Viruses

Barley yellow-dwarf virus (BYDV)
Maize dwarf mosaic virus (MDMV)
Tomato mosaic virus (TMV)
Potato virus diseases
Wheat streak mosaic virus (WSMV)

Diseases Caused by Fungi

Club Root of Crucifers: *Plasmodiophora brassicae*

Club root attacks a wide range of cruciferous plants in the cooler regions of the temperate zones, Brussel sprouts, cabbage, cauliflower, rape, swedes and turnips being important economic hosts; but wallflowers and stocks as well as weeds such as charlock and shepherd's purse can also be attacked and can act as dangerous reservoirs of inoculum.

The main symptoms of root deformation by tumefaction at best lower the marketable quality of the crop, and at worst are forerunners of plant death. The fungus is a very primitive member of the Plasmodiophorales in the Myxomycetes and is a true soil-inhabiting organism. Infection takes place through root hairs and the first symptoms are a stunting, wilting and yellowing of the seedlings. Root examination will reveal the typical clubbing or gall formation on the roots (Compendium Plate 3(ii)), often producing an appearance which has resulted in the common name of 'finger-and-toe' disease on plants with a fibrous root system like cabbage or kale. In cases of severe infection the galls may decay leading to plant death, but significant reductions in yield will occur with all but the slightest level of infection.

The life-cycle of *P. brassicae* can be considered as starting with the germination of the resting spores which can survive in soil for at least seven years with a twenty year

The life-cycle of *P. brassicae* can be considered as starting with the germination of the resting spores which can survive in soil for at least seven years with a twenty year longevity being considered the extreme. However, in the warmer soils of the southern states in the United States, no spores survive the winter as they germinate in the moist, warm soils.

On germination, the resting spore releases a single biflagellate zoospore which can infect root hairs of susceptible plants after losing or withdrawing its flagella. Once inside the plant, the fungus produces an amoeboid thallus, or plasmodium (no hyphae are produced) which differentiates into zoosporangia. The next step is intriguing in that the zoosporangia then liberate four to eight zoospores each of which return to the soil surrounding the root system. It is then believed that the zoospores fuse in pairs (presumably of different but compatible mating type) to form a diploid zygote which can reinfect the root. Infection of the cortical cells by zygotes result in a diploid thallus, a plasmodium, being formed which, by invading and colonizing the cortical tissue, stimulates extensive **hyperplasia** and **hypertrophy** to produce the clubbed roots. Ultimately, and presumably following meiosis, the plasmodia differentiate to produce masses of small, round, haploid resting spores which pack the infected cells and which are released into the soil on the decay of the roots.

The pathogen is most active in soil between 20 and 25°C and infection is favoured by wet, acid soils. Dispersal is easily effected by the spread of contaminated soil on implements, the feet of animals and man and around the roots of transplants. Many physiological races of *P. brassicae* have been identified and they generally occur in mixed populations.

Club root can be confused with other gall formations on cruciferous plants. Some swede cultivars produce swellings on the roots which contain no parasite and apparently do no physiological damage. These are called **hybridization nodules**. A most common gall symptom is produced as a result of invasion by the **turnip gall weevil**. In these very rounded galls, a maggot can often be found in the central cavity.

Control

With inoculum being produced in large amounts in gall tissue, great care should be taken over the disposal of infected plants. Preferably, such roots should either be destroyed or fed to animals on the spot as resting spores can pass through the animal and could be spread in the manure. Care should also be taken to eliminate the transport of soil from contaminated areas on implements or feet and hooves. Transplants should always be raised in soil with no history of the disease and the lime status of the soil where the crop is to be grown should be adjusted, preferably to a level of pH 7.0 or above as the spores will not germinate at this level of alkalinity. Soil moisture levels should be reduced by draining, although heavy rain in the late summer can soon restore levels to above the minimum threshold for infection of about 50% moisture-holding capacity.

There is no satisfactory fungicide treatment for drilled crops although treating transplants with a dip of mercurous chloride (calomel) has proved very successful although this chemical is prohibited in some countries. The systemic MBC compound, thiophanate-methyl, can also be used as a dip and there are claims for other systemic chemicals but none is as effective as calomel.

Cultivar resistance has been incorporated in forage rape, swede and turnip but only to certain races of the pathogen. There is very little resistance in broccoli, Brussel sprouts, cabbage and cauliflower.

Powdery Scab of Potatoes: *Spongospora subterranea*

This is one of the two potato scab diseases, the other being a bacterial disease, common scab, caused by *Streptomyces scabies*. It is caused by a soil-borne fungus which was first described in the United Kingdom in 1846, and was considered to have been introduced to Europe from Peru where it is endemic. It has now been reported in almost all potato-growing countries.

The disease is only serious under conditions of high soil moisture coupled with relatively low soil temperatures (< 14 °C) and is most severe in northern areas of the United Kingdom and the north-eastern states of the United States. Liming increases powdery scab (as well as common scab) but provided favourable temperature and moisture conditions prevail, variation in pH over a wide range (4.6–7.6) appears to have little effect. The fungus can infect the roots, stems, stolons and tubers but it is only the tuber infection which is of significance. The first symptom on the tubers are small raised brown swellings which might grow to about 6 mm in diameter in the first week after infection (Plate 3.1(i)). Their size is restricted by the formation of a corky layer of cells which forms immediately below the scab lesion. The scabs are generally circular in shape and can often be confused with the common scab disease. They may be isolated or they may coalesce to produce a vary large scabbed area. Eventually, the epidermis ruptures and a mass of dry, powdery brown spores are exposed. These spores, which in fact are grouped together to form characteristic composite spore-balls each of which is enclosed in a transparent membrane, are the resting stages of the fungus and they can be released into the soil or contaminate healthy tubers when they are lifted, the disease being thus both soil-borne and seed-borne.

The resting spores germinate to release a single biflagellate zoospore which infects the host through a root hair. After penetration, it develops a primitive thallus which differentiates to produce about 50 zoospores which are released back into the soil. These reinfect the host through lenticels or by direct penetration initiating further cycles of zoospore production and infection under favourable conditions. If two zoospores fuse in the soil, the amoeboid product can also infect the host where it produces a plasmodium which develops intercellularly and from which, presumably after a meiotic division, spore-balls containing resting spores are produced in the host cells.

Under very wet conditions, more severe symptoms may be produced, the canker form on the tubers. Here, the presence of the pathogen induces secondary growth leading to grotesquely misshapen tubers which are often mistakenly diagnosed as wart disease symptoms.

On the roots of the potato plant the infections often lead to a stimulation of host cell growth with the production of small outgrowths or galls which also produce masses of spore-balls which can remain in the soil for at least three years depending upon environmental conditions. These spores have another role in potato disease epidemiology for it is now known that they can act as the vector for potato mop top virus.

Control

With the soil often providing the necessary inoculum for infection, a suitable rotation of at least three years free of potatoes is advisable. The movement of contaminated soil or compost should also be strictly controlled. Drainage will also help to reduce disease incidence but irrigation, whilst a common practice in potato cultivation and one which will reduce common scab infection, tends to increase powdery scab.

Levels of inoculum on seed tubers can be reduced by mercury dipping. However, this treatment is seldom practised in the United Kingdom as the disease rarely becomes sufficiently severe. Care should be taken not to plant any scabbed seed tubers and it is an offence for such infected potatoes to be sold. No cultivars of potatoes are completely resistant to powdery scab although there is a considerable range of susceptibility.

Wart Disease of Potato: *Synchytrium endobioticum*

Early in the twentieth century, wart disease seriously threatened the future of the potato crop in the United Kingdom. The use of immune varieties coupled with rigorous legislation have now relegated this disease to being of infrequent occurrence and causing negligible losses nationally. However, the potential threat of this disease has not diminished and our present control strategies should never be weakened.

The fungus is soil-borne, infecting the tubers and stolons but not the roots and having little effect on the haulms. The tubers may be converted into grotesque, warted masses or they may produce a warty protruberance from one of the eyes, the size of which is often greater than the tuber itself. Not all the tubers on a plant are necessarily infected and there is often a wide range of sizes of the warty outgrowths on those that are attacked.

The causal organism is a very primitive fungus producing no mycelium, the body or thallus being represented by two spore stages, a thin-walled summer sporangium and a thick-walled overwintering resting sporangium both of which, on germination, release zoospores whose motility is greatly influenced by soil moisture which results in a greater severity of wart disease in wet seasons.

On lifting the tubers, it is inevitable that pieces of the warted material containing the resting sporangia will remain in the soil. These spores have a viability of up to thirty years and not only provide inoculum for subsequently planted potatoes but also can easily be transported in soil on boots and implements to contaminate clean soil elsewhere.

Control

Some cultivars are immune to wart disease and provide the only means of control. The resistance of such cultivars has proved to be long-lasting due mainly to the low dispersal capability of the pathogen, any new races of the fungus being restricted by lack of mobility to that area of soil where they were produced.

The disease is notifiable under European Economic Community legislation, a sanitation zone being demarcated around the contaminated field, and in which the planting of immune cultivars of potatoes only are allowed. The strict enforcement of these legislative orders have largely eliminated this disease as a major threat and it is important that no relaxation of these procedures is allowed.

In many countries various degrees of quarantine have been practised to reduce the pathogen spread, and in the United States the entry of potatoes is prohibited from any country where wart disease is known to occur.

Watery Wound Rot of Potatoes: *Pythium ultimum*

Pythium ultimum is an ubiquitous soil-borne pathogen of potatoes but it can only enter through wounds, albeit small wounds, of which the majority occur during lifting. The disease is thus mainly a problem of potatoes in store. The fungus has a high optimum temperature for growth and thus it produces its greatest damage if temperatures rise under storage. Conversely, there is little damage when temperatures are low.

The internal symptoms are unmistakeable with a watery, almost black rot in which cavities develop. On cutting open the tuber a distinct fishy odour is released. The rot usually continues to develop until the tuber flesh completely collapses and is converted to a dark liquid.

Control

Unfortunately, there is no control for this disease. The pathogen, being soil-borne, is almost always present often in the form of thick-walled resting spores, oospores. However, by endeavouring to minimize injury, especially during lifting, the incidence of the disease can be reduced. It has also been shown that by destroying the haulms of early crops about two weeks before lifting, the tubers become more mature and less liable to damage. Infected tubers, whenever they are found, should be gathered and destroyed as should any rooted tubers growing after harvest.

Late Blight of Potatoes and Tomatoes: *Phytophthora infestans*

The potato famine in the mid-nineteenth century in Ireland gave notoriety to the later identified pathogen which is still the most serious fungal invader of potatoes. The disease probably originated in Mexico gradually spreading into South America and eventually to

Europe and the United States between 1830 and 1840. The fungus, which is a member of the Pythiaceae in the Peronosporales of the Mastigomycotina, one of the downy mildews, attacks both haulms and tubers, losses of 50% being common when the foliage is destroyed by the end of July. The disease is also a serious problem in the outdoor tomato crop but is rarely a problem under glass.

The disease is very weather-dependent causing little or no damage in dry seasons but capable of rapid destruction in the haulms in wet seasons or seasons with frequent warm moist periods. Early potatoes usually are lifted before conditions become sufficiently conducive for the development of blight and it is the second-earlies and main crop that become most affected.

Initially, the disease can be observed as a few, scattered infected plants, or **primary foci**, with a small number of dark, water-soaked spots on the leaves. Under optimal conditions, the disease spreads rapidly from plant to plant and progressively over the total foliage of the individual plant, eventually killing off all the stems and leaves. The leaf symptoms are very similar in tomatoes. An examination of the undersurface of the necrotic leaf lesions would show a delicate white mould which is the asexual stage of the fungus (Compendium Plate 1(i)), the sporangiophores and characteristic lemon-shaped sporangia which are borne singly at the ends of these branched structures (Plate 2.1(ii)). The dispersal of these spores under rainy or humid conditions from plant to plant is primarily in the direction of the prevailing wind but, as the crop canopy closes between the rows, air currents can also spread the spores along the rows by the 'tunnel effect'. Rain is also responsible for washing the spores into the soil where they may infect the tubers especially if they are insufficiently covered by soil. The fungus is heterothallic, the sexual stage only appearing to occur in Mexico, the country of origin. In most other countries, only one mating type is present restricting the life-cycle to the asexual form. However, both mating types have recently been reported in the United Kingdom.

Infection of tubers can also occur at lifting if there are still partly blighted haulms present at this time. The external symptoms on the tubers are irregular, slightly sunken areas often giving a dark or purple discoloration. A slice through such an infected area shows a rusty, or reddish-brown dry rot (Compendium Plate 1(ii)) which may later become a wet rot if the infected tuber is invaded by secondary rotting organisms, especially bacterial soft rot (*Erwinia carotovora* subsp. *carotovora*, see p. 160). Tomato fruit are also attacked with olivaceous to brown areas appearing and the firmness quickly deteriorating to a completely rotten state.

The disease cycle can start as a result of planting infected tubers, from spores developing on piles of discarded tubers, from groundkeepers and, to a limited extent, from spread of spores from other infected crops in the locality. It is extremely difficult to recognize slightly infected seed tubers and these primary foci of infection regularly occur in crops.

The weather dependency of this disease can be used to advantage in that, with the help of meteorological data, it is now possible to give about two weeks warning of the outbreak of disease in most areas. Disease forecasts, which enable a more efficient timing of application of protective fungicide sprays, are available from the Agricultural Development and Advisory Services in the United Kingdom, in the press and on both radio and television. In some countries, computerized forecasting is available, the farmer telephoning meteorological and disease data to a central control. The Blitecast system in the United States is a good example (see p. 103). The conditions upon which the disease forecasts depend were originally as described by the so-called Beaumont period (see p. 103) which needed 48 hours during which the temperature does not fall below 10 °C nor the relative humidity of the air below 75%. Since 1975, a more precise definition of a 'blight infection period' has been used. This is defined as two consecutive periods (1300–1200 hours) in which the minimum temperature is 10 °C or above, and in each of which there are at least 11 hours with the relative humidity above 89%.

Control

Quite obviously, the planting of infected tubers should be avoided. Earthing up the rows will also reduce the possibility of tuber infection as will haulm destruction just prior to harvest. The current practice is to destroy the haulm two to three weeks before lifting although this timing will depend up on the susceptibility of the variety, the nature of the soil and the state of the ridges, and the weight of the crop already formed. Sulphuric acid is an excellent chemical for this purpose (70% H_2SO_4 at 225 l/ha) but it is very corrosive. Diquat and dinoseb are also used but only if the soil is wet.

Protective spraying was traditionally carried out using the copper-based Bordeaux or Burgundy mixtures, but these have been largely superseded by the organic dithiocarbamates like zineb, maneb and mancozeb and the organotins like fentin hydroxide and fentin acetate which also give some control of the tuber blight stage. Captafol, chlorothalonil and, very recently, the systemic metalaxyl are also used but fungal resistance to the latter has occurred recently. Protective spraying requires a series of interval sprays due to wash-off and new growth of foliage. The evidence points to there being little difference in efficiency in the manner of spraying, be it high volume (1100–1350 l/ha) or low volume (225–280 l/ha).

Chemicals can also be used for the control of sporulation from discarded piles of potatoes. Desiccant herbicides, persistent herbicides or even sodium chlorate can be used for this purpose.

Potato cultivars vary in their susceptibility to blight. All early varieties are regarded as susceptible and, whilst they are rarely seriously affected because of the time of their lifting, they can provide an important inoculum source for the main crop. Some main crop varieties have resistance to haulm blight only, while others show some resistance to both haulm and tuber blight.

Downy Mildew of Tobacco: *Peronospora tabacina*

Downy mildew of tobacco, sometimes called blue mould, is an important disease in almost all the tobacco-growing regions of the world but especially Australia, the Middle East and the United States causing considerable losses in plants in the seed beds. In 1959, it spread throughout the whole of Europe with losses reaching 80% in some tobacco-growing centres.

When young bed plants are infected, the first symptoms are a cupping and twisting of the diseased leaves, the tips of which turn yellow. A spotting with somewhat indistinct margins appears on the upper leaf surface. The downy mildew felt of blue-tinged mycelium may be seen on the lower surfaces. These blight symptoms are limited by the veins and the necrotic spots often fall out of the leaf. Extensive blighting is accompanied by a rank odour and plant death is common. Field-grown plants rarely develop blight symptoms although the leaf spots eventually become necrotic.

The causal organism is a typical member of the Peronosporales in the Mastigomycotina. It has both asexual and sexual stages with the sporangia produced on the characteristic claw-shaped tip of the dichotomously branded sporangiophores which emerge through stomata. These wind-borne sporangia are the cause of the successive secondary cycles of infection which are favoured by cool, wet weather. Sporangia are produced optimally at 13–15 °C and also germinate best at around these temperatures as long as there is a film of moisture on the host surface. In fact, warm dry weather will check spore production and disease increase.

In the United States, sporangia move northwards in the prevailing winds, initiating successive cycles of the pathogen during the late winter and early spring. This predictability of spore movement, coupled with a knowledge of temperature effects, has led to a forecasting system based on weather conditions during the intercrop period. The severity of blue mould in the southern United States is related to winter temperatures, especially those in January. Probably by destroying oospores or the secondary hosts, low January temperatures normally give rise to low levels of disease in the growing season and retard the onset of disease. Spores from the tobacco areas of the United States are thought to cause infection in Canada about 2–3 weeks after being produced. Conducive January temperatures would give high disease levels in the United States but also give ample warning of possible epidemics in Canada.

The sexual stage, with the production of thick-walled oospores in diseased tobacco leaves, provides the pathogen with a means of overwintering although various *Nicotiana* species can also harbour *P. tabacina* and produce conidia which provides inoculum for the primary infection cycles.

Control

P. tabacina control involves an integrated programme, with cultural measures designed to improve the aeration of the crop by increasing the planting distance and subsequently picking the lower leaves, being combined with sanitation procedures to destroy crop debris. The use of new land for seed beds to avoid overwintering oospores is highly recommended but eradicative practices may be employed such as soil fumigation with methyl bromide before seeding.

The use of chemical foliar sprays is aimed at protection against incoming spore clouds. Regular application of dithiocarbamates such as maneb, zineb or mancozeb, often as dusts, are recommended at weekly intervals during cool, wet weather. Resistance is available in *Nicotiana debneyi*, a related species and breeding programmes are introducing this resistance into commercial cultivars.

Other *Peronospora* spp. include *P. destructor*, the onion downy mildew which can cause extensive damage in wet seasons. Leaf collapse with the associated downy mildew felt, the asexual stage, typifies the disease during the growing season. Bulbs from an infected crop become soft and shrivelled after a period of storage. Oospores may be produced providing a primary source of inoculum over their 4 to 5 year period of longevity. *P. parasitica* attacks many plants in the Cruciferae. It is common on many economic Brassicae spp. as well as bedding plants like stocks, alyssum and wallflowers. It also attacks cruciferous weeds such as shepherd's purse (*Capsella bursa-pastoris*) which can provide sources of inoculum for the cultivated species.

Lettuce Downy Mildew: *Bremia lactucae*

Downy mildew of lettuce is frequently encountered by vegetable growers all over the world but it is especially important in the north temperate zone where moisture is adequate and the temperatures none too high. The disease may occur at any time during the growing season although the symptoms in the young seedlings may not be apparent, usually just a slight yellowing of patches of the leaves, between the veins.

On the older leaves, the disease is more easily recognized by these pale angular patches which either rot in wet weather or, if dry, turn brown and papery. Corresponding to these infected areas, on the lower leaf surface the fungus can be seen with the naked eye as a downy white superficial covering. This growth, which can sometimes appear on the upper surface, is the asexual sporulating stage of the fungus, masses of sporangiophores and sporangia which protrude out of the many stomata on the leaf surface. For domestic purposes, trimming away these infected leaves will often still leave satisfactorily sized lettuce but such lettuces are useless for commercial sale.

Downy mildew can occur at any time but is most severe in early spring and late summer crops outdoors and in the glasshouse crop grown with heat in the winter. The disease can spread from plant to plant by airborne sporangia released from the downy growth on the infected leaf surface. These spores can only infect the leaf if it is wet and, when they germinate, they produce either the tube-like outgrowths called germ-tubes or release motile spores which can swim in the film of moisture. The cycle is very quickly repeated for, within 5–9 days of infecting the leaf, the fungus will again be producing spores on the leaf surface.

Like most downy mildew pathogens *Bremia* is very sensitive to temperature, moisture and light hence sporulation occurs mainly at night when the night temperatures are fairly low, between 6 °C and 10 °C, and the humidity very high. Such conditions confer special importance to this disease in the early autumn on outdoor lettuce and in late autumn on frame lettuce.

The fungus can also reproduce sexually to produce resting spores, oospores, which are released into the soil from infected leaves when these rot away. These can overwinter in the soil and infect the next lettuce crop. However, sexual reproduction is seemingly a rare event and most of the inoculum comes from infected lettuce debris and even wild lettuce, the spores often being blown long distances in the wind.

Control

The incidence of downy mildew can be reduced by good husbandry. In the glasshouse, it is possible to minimize the time during which the leaves are wet by not watering too late in the evening. Drying out of the leaf surface will also be hastened by good ventilation in both glasshouses and under cloches, and even the removal of windbreaks in the field situation will help in this context. A rotation of crops is always advisable especially if there has been a history of downy mildew when lettuce should not be grown in contaminated areas for several years. Infected plants should also be destroyed and lettuce debris collected and similarly treated.

There are no cultivars of lettuce which are universally resistant due to the fact that many races of the fungus occur and are distributed in varying proportions in different geographical areas. Certain cultivars have had resistance genes bred into them which give them resistance against some but rarely all the races in a particular locality. The grower can only be advised to look for the best variety for his area by trial and error of the many cultivars commercially available.

Spraying is not normally a practice readily acceptable to the consumer when the fungicide is directly applied to the edible produce. Many compounds have been used to reduce downy mildew incidence, the dithiocarbamate Dithane in particular. This has now been superseded by the newer compound metalaxyl which is both safe and very effective but must be applied at regular intervals and stopped at least 2 weeks before harvesting. However, the regular use of such systemic compounds might result in strains of the pathogen becoming resistant to the fungicide as has happened in other disease situations.

Peach Leaf Curl: *Taphrina deformans*

This disease has been recorded in Europe and the United States since early in the nineteenth century. It is of economic importance wherever peaches are grown but especially in the Great Lakes region of the United States, and the western valleys of Washington, Oregon and California.

The pathogen is a member of the Taphrinales in the Ascomycotina and attacks both leaves and twigs, especially after a cool moist spring. In addition to peaches, almonds, nectarines and occasionally apricots can be attacked and related species can be seen on other stone fruits and flowering *Prunus* trees.

The first symptoms appear as the young leaves begin to unfold in the spring. Infected leaves show a puckering and a reddish coloration the extent of which varies from a small blister-like distortion to a curling of the whole leaf (Compendium Plate 3(i)). The curled areas later appear silvery due to the development of a waxy bloom. Badly affected leaves die and drop prematurely. A section through the leaf at the silvery stage would reveal an extensive intercellular mycelium which produces rows of asci between the epidermis and the cuticle, the cause of the waxy sheen. The asci are naked with no ascocarp being produced. Usually eight unicellular ascospores are formed in each ascus but these can bud off secondary spores, or conidia, so that ripe asci may be packed with spores.

On discharge, the ascospores can become lodged on the surface of twigs and between bud scales where they may overwinter. The spores germinate in the spring as the bud opens, producing germ-tubes which penetrate the young leaves directly through the cuticle. Infection by a single ascospore has been shown to give rise to the formation of the sexual stage, the asci and ascospores, indicating that the fungus is homothallic.

The fungus not only attacks the leaves but also invades the young shoots and, more rarely, the flowers and fruits. Infected young shoots become swollen and twisted, their diseased leaves usually forming a tuft on the ends of stunted shoots. Young fruits might also drop prematurely.

Control

Satisfactory control of peach leaf curl can be achieved by the single application of a protectant fungicide before bud-break in the spring. Traditionally, Bordeaux mixture (1 kg $CuSO_4$ and 1.25 kg $Ca(OH)_2$ in 100 litres water) has been used but 3% lime sulphur is a good alternative as are several of the organic fungicides of the dithiocarbamate group. Removal of affected leaves before the waxy bloom develops should also be undertaken routinely, but only as a component of a programme which includes fungicide application in the spring.

Cereal Powdery Mildew: *Erysiphe graminis*

This is undoubtedly one of the most important diseases of cereals, especially barley and wheat. It is particularly prevalent in regions where humid conditions prevail during the growing season. In Canada, British Columbia and the eastern provinces are high risk areas as is the whole of the United Kingdom and most of Europe; and in the United States it is most severe in the Pacific coastal regions, around the Great Lakes and on the Atlantic seaboard.

The disease can be seen on leaves, sheaths, stems and inflorescences. In the first instance, some yellowing and curling of the leaves may be seen and close inspection will reveal superficial mats of white to grey mycelium on the aerial plant parts. Microscopic examination of this stage will reveal septate mycelium from which masses of conidia are produced in long chains on short bulbous conidiophores (Plate 2.1(i)). The most mature conidium is distal on the chain and is easily detached giving a powdery appearance to the mycelial mat.

These mildewed areas enlarge, coalesce and darken and the sexual stage can be seen in the appearance of dark brown cleistothecia embedded in the mycelium. The cleistothecia have simple, myceloid appendages and contain 9–30 asci, each containing eight ellipsoid ascospores.

Infection can occur after the deposition of either conidia or ascospores on the host tissue, both directly penetrating the cuticle. The conidia usually germinate best over a

(i)

(ii)

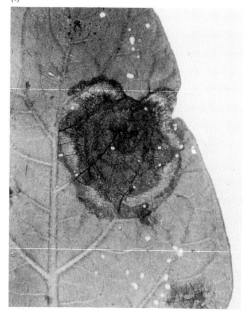

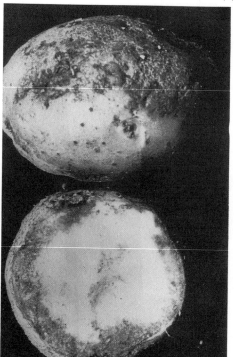

(iii)

Plate 1
   (i) Late blight of potato (*Phytophthora infestans*). Note white halo of emerging sporangiophores around necrotic lesion
   (ii) External and internal symptoms of late blight on potato tubers
   (iii) Take-all of wheat (*Gaeumannomyces graminis*). Note blackened and sparse rooting system
   (iv) Apple scab (*Venturia inaequalis*). Note superficial scabs and cracking.

(iv)

Plate 2

   (i) Brown rot (*Sclerotinia fructigena*) an apple showing conidial sporodochia

 (ii) Raspberry rust (*Phragmidiums rubiidaei*) urediosori on leaves

(iii) Choke (*Epichloe typhina*) on grass. Note the light coloured fungal stroma prohibiting the emergence of the inflorescence

(iv) Leaf blotch (*Rhynchosporium secalis*) on a barley leaf.

(Photos: (i) & (ii) Shell Photographic Unit)

(i)

(ii)

(iii)

(iv)

Plate 3

  (i) Peach leaf curl (*Taphrina deformans*)
  (ii) Club root (*Plasmodiophora brassicae*) on swede
 (iii) Loose smut (*Ustilago nuda*) on barley
  (iv) Ergot (*Claviceps purpurea*) on rye.
(Photos: (i) Shell Photographic Unit)

(i)   (ii)

(iii)

Plate 4

  (i) Potato plant showing advanced virus leaf roll symptoms

 (ii) Section through a potato tuber infected with bacterial ring rot (*Corynebacterium sepedonicum*)

(iii) Aecidial stage of *Pucinia graminis* on the alternate host, barberry.

range of temperature from about 10–15 °C and at a relative humidity of 95%. However, conidia have an unusually high water content and germination can occur even at zero humidity. On germination, the conidium produces a short germ-tube the end of which swells out to form an **appressorium**. Penetration of the cuticle is effected by a slender penetration peg which emerges from the lower appressorium surface. Once in the epidermal cell, the penetration hypha develops a rooting and anchorage structure, an elliptical body with finger-like appendages, the **haustorium**. The haustorium is the only internal structure, providing the necessary sustenance to allow the development of the superficial growth of mycelium and sporing bodies. Masses of conidia can be produced, providing ample wind-borne inocula to rapidly cause disease levels of epidemic proportions. Conidia can be spread long distances by wind although they are not generally longlived.

Ascospores are discharged mainly in the autumn after the inbibition of water by the cleistothecium. The ascospores can infect autumn-sown cereal crops or certain grasses. Cleistothecia may survive on stubble debris over the winter, discharging their ascospores in the spring.

Infection produces a rapid chlorosis and necrosis and, with early infections can reduce root growth and tiller numbers. Infection of the glumes is quite common on wheat and can affect grain size.

Erysiphe graminis is a member of the Erysiphales in the Plectomycetes of the Asco-mycotina and it exists in several forms, or *formae speciales*, distinguishable by the hosts attacked. *E. graminis* f. sp. *tritici* is the form which attacks wheat whilst barley is attacked by *E. graminis* f. sp. *hordei*. However, these forms are not totally host-specific as a certain amount of promiscuity does exist. In addition, within each *formae speciales*, physiological races exist which can be differentiated by testing on a set of host cultivars each one possessing a different resistance gene. The distribution and prevalence of these virulences, or physiological races, is monitored in many countries, the survey information being of immense value to plant breeders.

Control

With a high probability of inoculum overwintering on stubble debris or on volunteer plants, cultural methods of control involving either ploughing or herbicide application are recommended. Much of the inoculum for the spring-sown crop originates from the autumn crop and attempts to keep the latter free of disease by the use of chemical seed treatments have produced delays to the epidemic in summer although, because of the long-distance transport of spores, such treatments are not likely to be very successful.

The use of race-specific resistant cultivars has been the main method of control since the 1930s. This has been reasonably successful although the advent of new races within about four years of the introduction of a cultivar possessing a new resistance gene, has led to the **boom-and-bust** cycle (Figure 6.3) with a rapid turnover of cultivars. Diversification schemes have been introduced into several countries. In the United Kingdom, the cultivars are grouped according to their resistance genes and the farmer is advised to use a selection of cultivars; each one used to be chosen from a different diversification group (Table 6.2). Alternative strategies of diversification involve the use of **multigene** or **multiline** cultivars or cultivar **mixtures**. This approach to disease control has been discussed in Chapter 8. Many modern cultivars combine race-specific resistance with the more durable but less effective 'field resistance' which is race non-specific. Should the specific resistance be overcome, then the plant still has the background resistance for protection.

The chemical control of cereal mildew has only been practised on a wide-scale since the early 1970s. Now, with many systemic or partly systemic chemicals available, chemical control can be effected with seed treatments, foliar sprays or, in many cases, the use of both methods of application. For winter barley, autumn and/or spring and early summer fungicide applications may be used. For spring barley, the decision to apply a foliar spray will normally be made on the basis of the likely disease risk. As an alternative to autumn spraying, seed treatments may be applied, ethirimol, flutriafol, triadimenol and triforine being widely used. Foliar sprays include fenpropimorph, prochloraz, propiconazole, thiophanate methyl, triadimefon, tridemorph and triforine and there has also been a movement towards the use of fungicide mixtures, for example tridemorph and carbendazim, this approach being adopted in the fear that the pathogen might develop strains which are insensitive to the more site-specific chemicals. These have been found but have not multiplied in the population to the extent that fungicide efficiency has been reduced significantly. The timing of foliar sprays is most important. In the autumn, spraying is usually recommended if there is between 5 and 10% mildew on the leaves. In the spring, use is made of the forecast system based on high-risk periods (see p. 104) but, as a rule of thumb, it is advisable to spray as soon as mildew appears on the new growth. In the late spring and summer, spraying is recommended if mildew affects 3% of the second or third leaf before ear emergence.

Other Powdery Mildews

The Erysiphales contains many pathogens of economic importance. *Sphaerotheca morsuvae* causes the American powdery mildew of cereals where both leaves and fruit are attacked. Both asexual and sexual stages are produced, the latter being very obvious as a dark green/brown, superficial mycelial felt over the fruit in which numerous cleistothecia, each containing a single ascus, are embedded. *Microsphaera grossulariae* causes the European powdery mildew of gooseberries but it should be noted that it is *S. mors-uvae* that is most damaging in the United Kingdom. The cleistothecia of *M. grossulariae* contain several asci but can be distinguished microscopically by the dichotomously branching ends to the appendages. In contrast, *Podosphaera leucotricha*, the causal organism of apple powdery mildew, only produces a single ascus although possessing dichotomously branched appendages on the cleistothecia. *Uncinula necator* causes powdery mildew of vines, the cleistothecia being characterized by appendages with single-hooked or double-hooked ends. *Phyllactinia corylea* causes powdery mildew of deciduous trees and is easily recognizable microscopically by the unique, bulbous-based and elongated spikes of the cleistothecial appendages. Most of the powdery mildews can be controlled by sulphur-based fungicides, but there are many systemic fungicides now available that are equally as effective.

Apple and Pear Scab: *Venturia inaequalis* (*V. pirina*)

Scab is the most important disease affecting the aerial parts of apple and pear trees. It attacks leaves, twigs and fruits, the latter often being so badly affected that they are of no economic value. It is a very good example of a quality-reducing disease. Affected leaves cannot function efficiently and attacks on young shoots which occur on certain cultivars can open the way to infection by other pathogens.

The first symptoms appear on the lower sides of young leaves where small, dull, olive-green patches may be seen. Later, lesions may also be seen on the upper leaves. In moist weather, the lesions become 'felty' due to the production of conidia which are borne singly at the tops of very short, slightly wavy conidiophores. Infection of the twigs is common in the United Kingdom but rare in the United States, the symptoms being areas of cracked, blister-like eruptions in the bark. Scab spots on the fruit may follow an early infection of the sepals. The scabs may be small and scattered or there may be extensive scabbed areas, often with an associated deformation of fruit shape and which develop deep cracks which will allow secondary invaders easy access (Compendium Plate 1(iv)).

The sexual stage occurs in the fallen leaves on the orchard floor during the winter, mature pseudoperithecia being produced in which asci each containing eight two-celled ascospores are produced. When fully mature, the asci will take up water and, as hydrostatic pressure builds up, the ascospores are forcibly ejected through the perithecial ostiole. Ejection only forces the spores about 2 cm upwards from the decaying leaf surface, but this is sufficient for wind currents to carry them up to the level of the foliage. The twigs also provide spring inoculum in the form of conidia which can be dispersed by wind or rain-splash.

Both conidia and ascospores will infect given the correct conditions. The optimum conditions for infection have been described and have been used for the forecasting of apple scab epidemics (Chapter 8). The fungus belongs to the Loculoascomycetes in the Ascomycotina.

Control

The initial aim of any programme for scab control should be to reduce the overwintering inoculum. This can be achieved by gathering up and burning the leaves, spraying the leaves on the orchard floor with an eradicative fungicide such as DNOC (dinitro-orthocresol) or by applying 5% urea which encourages leaf decomposition.

Eradicative chemicals like phenylmercuric chloride applied in the autumn have also given good control of the inoculum in fallen leaves as well as in twigs. The pruning of infected twigs is also recommended.

The chief control measure is protective spraying. Traditionally, this was carried out at three growth stages, the 'green bud' stage, the 'pink' stage when the petals are showing but have not yet opened, and the 'petal fall' stage. Several fungicides have proved successful in this routine spray programme, captan, dodine-acetate, maneb and wettable sulphur being good examples. Lime sulphur is probably the oldest chemical for scab control but certain cultivars may become scorched and suffer premature leaf drop. Organomercury compounds, although not allowed in some countries, have also given good control and have the advantage of giving a curative effect if applied within 72 hours of infection. The timing of application will greatly affect the efficiency of spray programmes and the use of the forecasting criteria—the Mills period, or the Smith period (see Chapter 8)—are of particular relevance in this respect.

Some species of *Malus* are resistant to scab, the resistance being governed by multiple genes. Breeding programmes are in existence but, with the pathogen having several physiological races, the long-term future for resistant cultivars is in some doubt.

Ergot of Cereals and Grasses: *Claviceps purpurea*

Ergot is a disease of a wide range of cereals and grasses with rye being very susceptible and, in descending order of susceptibility, wheat, barley and oats being more resistant. It affects only the flowering parts of the host plant, producing sclerotia (ergot) which completely replace the grain in the ears. These hard, black ergots may fall to the ground before harvest or, in some cases, they may be harvested and either sown again with the seed or, possibly, eaten directly by animals. They might also be ground up along with the grain and contaminate the flour or cornmeal. Unfortunately, ergots contain noxious substances, mainly alkaloid materials, which are injurious to livestock and to humans if contaminated flour is eaten. There appears to have been only one occurrence of typical human gangrenous ergotism in the United Kingdom and that was as far back as 1762 in Bury St Edmunds, but other similar tragedies have been reported in Europe and other countries in recent times. The toxins can also produce severe convulsions in man and animals. Drugs prepared from ergot extracts have medicinal value and are used in surgical operations where blood vessels require dilatation.

Ergots can remain in the resting state until the following year when they germinate, if they are on or near the soil surface, at about the time the cereal crops come into flower. Ergot germination is influenced greatly by temperature and it has been shown that ergots exposed to temperatures below freezing for 1 month will produce stromatal heads most rapidly at 9–15 °C and are inhibited at 18 °C and above. As many as sixty stromatal heads may arise from each ergot, the perithecia being completely sunken in the head. The ascospores are extruded from the perithecial ostiole in a viscous fluid under conditions of high air humidity (76–78% relative humidity). Spread of the ascospores to neighbouring flowers is by insects, rain-splash and to some extent, by wind. The susceptibility of various hosts may be related to the length of time the flowers remain open. This feature is of prime concern to the plant breeders who are at present engaged, in several cereal crops, in the breeding of hybrid varieties. The florets of the male-sterile lines of self-pollinated cereals are open for a long time and are thus susceptible for an extended period. Disease incidence of over 76% of the heads and 36% of the florets has been reported in male-sterile barley in the United States (Puranik and Mathre, 1971).

The ascospores germinate on the stigmas and soon infect the developing ovaries. The conidial stage is produced as numerous short conidiophores become closely packed over the convoluted surface of the remains of the ovary. Small, unicellular, hyaline conidia are produced at the tips of the conidiophores, being abstricted in succession and embedded in a sticky fluid known as 'honeydew' which attracts insects which passively carry the spores to other healthy florets. The generally accepted view is that these conidia germinate to produce mycelium which penetrates the ovary near its base. The mycelium gradually ramifies in the enlarging ovary and, instead of grain, a sclerotium is eventually produced which projects out of the floret to a length often three times that of a normal grain but the size varies according to the host (Compendium Plate 3(iv)). The fungus is a member of the Hypocreales, in the Pyrenomycetes of the Ascomycotina. The severity of this disease is favoured by cool, wet weather during the flowering period. Cross-infection studies have shown that wheat, barley and rye could be infected with fungal isolates from thirty-eight different graminaceous host species (Campbell, 1957 and Mantle and Shaw, 1977) have found that *Alopecurus myosuroides* and *A. pratensis* have important roles in the epidemiology of the ergot disease of cereals.

Control

The disease rarely achieves economic importance and it is not usually necessary to resort to control measures. Varietal resistance has been shown to differ and pathogenic races are known to exist. The use of ergot-free seed is most important and this can be achieved by a flotation method using solutions of sodium chloride or potassium chloride or by screening. Deep ploughing and rotation of crops are effective remedies because ergots buried about 25 cm deep do not germinate and appear to rot away after a year. It has also been demonstrated that reasonable control of ergot on male-sterile barley can be achieved by the application of 2400 μg/ml benomyl applied three times prior to and during anthesis.

Take-all or Whiteheads of Cereals and Grasses: *Gaeumannomyces graminis*

Take-all disease has become of considerable importance in areas of intensive or continuous cereal cultivation. *Gaeumannomyces graminis* var. *tritici* is the form which attacks wheat, barley, several grasses, and rye, the latter less seriously, whilst oats are mostly immune.

G. graminis var. *avenae* attacks oats very readily and also wheat, barley and rye.

The appearance of patches in the crop where many young seedlings have been killed outright gives the first indication of this disease. These patches will eventually become infested with weeds. Some infected seedlings may survive but they will be stunted and produce fewer tillers. Such plants may eventually produce heads with empty spikelets and appear bleached. This symptom has given rise to the name 'whiteheads'.

The root system of infected plants is much reduced and there is blackening extending up to the lower leaf bases and the whole plant can easily be removed from the soil by gentle pulling (Compendium Plate 1(iii)). Perithecia are produced in this region, their necks protruding through the leaf bases with the main body being sunken in the stem tissue. Microscopically, a good diagnostic feature is the presence on the surface of the roots of thick, brown runner hyphae.

The production of perithecia, asci and ascospores is the only reproductive stage, the ascospores being ejected from the asci after a period of rain. However, ascospores are not thought to be of great importance in the infection process although they have been shown to be able to infect any seminal roots exposed above the soil surface. Infection takes place by means of slender hyphae which emerge from hyphopodia formed on the runner hyphae. The fungus does not grow systemically within the developing plant, death resulting from the restriction or cessation of the flow of nutrients and water. The fungus is a member of the Sphaeriales in the Pyrenomycetes of the Ascomycotina.

The fungus survives the intercrop period on stubble debris and this constitutes the main source of inoculum. The roots of autumn-sown crops may become infected during the winter but, mostly, infection occurs after May. The soil inoculum can rapidly build up and even as few as 1% infected plants in a crop can result in sufficient residual inoculum to produce severe take-all infection in the next cereal crop.

Take-all is most serious on light, loose alkaline soils of poor nutrient status where moisture is abundant. An abundance of soil nitrogen will enable plants to grow more strongly and to replace roots damaged by take-all even though it does not reduce the level of disease in the field in absolute terms. However, nitrogen applied the year before the cereal is sown will increase the survival time of the inoculum in the soil and thus increase the disease potential in the cereal crop.

An interesting feature of continuous cereal cultivation has been the decline of take-all after about the fourth year of cropping. Many explanations have been put forward to explain take-all decline, none have been positively confirmed but the phenomenon is unquestionable.

Take-all cannot be controlled by fungicide application. Rotation reduces the inoculum and even one year free of a susceptible crop will almost eliminate the potential for disease in the next cereal crop. An integrated approach should include a balanced fertilizer programme, the control of susceptible grass weeds and the early ploughing under of diseased stubble to encourage decomposition of the plant debris and the eventual starvation of the fungus. Seedbeds should be firm and well-drained. Conditions under grassland are relatively unfavourable to the take-all fungus compared with condition under a cereal crop. One of the reasons for this is the presence of highly competitive *Phialophora* species. It has been demonstrated that this fungus decreases the spread of *Gaeumannomyces graminis* along wheat roots. For this reason, direct drilling of wheat into a pasture after treatment with a herbicide like paraquat produces less take-all than in wheat drilled into cultivated land.

Some cereal varieties show a degree of take-all resistance, and the possibility of obtaining an increase in resistance in commercial varieties should not be overlooked.

Glume Blotch of Wheat and Barley: *Leptosphaeria nodorum*

Better known by the name given to the asexual stage of this pathogen, *Septoria nodorum*, this is one of the most important foliar diseases of wheat, especially winter wheat. Barley, rye and triticale are also attacked. Predominantly a leaf disease, the pathogen also attacks glumes, leaf sheaths and nodes. Early leaf symptoms are yellowish to tan-brown, oval or lens-shaped spots with rather darker borders. The leaf spots often coalesce to give extensive necrosis and leaf death. A toxin is produced which gives the diseased leaves a gingery-brown colour. Light brown pycnidia develop randomly in scattered colonies and are distinguishable from those formed by *S. tritici* which are more pronounced and black in colour. On barley, the symptoms are very similar but never as extensive.

Infection of the nodes can be severe, often causing stalk-break, the discoloured areas also bearing abundant pycnidia. The pathogen becomes more aggressive as the crop approaches maturity and head infection can be very severe. Infected heads are very conspicuous due to the discoloration of the glumes, a purplish-brown with a hint of grey, mostly developing from the tip downward (Plate 3.1(iv)). Pycnidia are also produced on the infected areas of the glumes and the seeds may also become infected. Severe infections can cause reductions in yield of up to 50% and a characteristic effect of this disease is the production of shrivelled grain.

The pathogen can thus be seed-borne but much inoculum can overwinter on stubble debris, volunteer plants and many grass species. The sexual stage (*Leptosphaeria nodorum*) has been found in several countries but the epidemiological importance of the ascospores produced has yet to be evaluated.

Sporulation from pycnidia extends over long periods, the pycnidiospores being extruded in a gelatinous, worm-like tendril called a cirrhus. The spores are splash-dispersed, initially about 1 m from the inoculum source and in the direction of the prevailing wind. The spores will germinate and infect between 5 and 30 °C in the presence of free moisture, the optimum being approximately 22 °C.

The fungus is classified in the Pleosporales of the Loculoascomycetes in the Ascomycotina and, although the symptoms are often confused with those induced by *S. tritici*, the latter has a different sexual stage (*Mycosphaerella graminicola*). Reference has already been made to the difference in pycnidia colour and microscopic examination of the pycnidiospores will also readily differentiate between the two species, *S. nodorum* having short (22–30 μm), mostly three septate spores, whereas *S. tritici* has filiform (43–70 μm), two to three septate and slightly curved spores.

Control

Recent work has shown that seed treatment with benomyl and triadimenol significantly reduced *S. nodorum* on the upper leaves and heads over a 3-year period (Luke, Barnett and Pfahler, 1986). Cultural methods of sanitation should also be practised and plant breeders have been fairly successful in achieving good levels of resistance in many present-day cultivars. However, resistance to *S. nodorum* is not linked to resistance to *S. tritici*. Several chemicals are now successfully used to control the leaf and head infections. The best control is often obtained when a spray is applied at or soon after flag-leaf emergence. Of the chemicals used, MBC, prochloroz, propiconazole and triadimefon are all highly recommended although in the United Kingdom there has been a recent increase in the numbers of *S. tritici* strains resistant to MBC fungicides and in such areas alternative chemicals should be used.

Net Blotch of Barley: *Pyrenophora teres*

Although primarily a foliar disease of barley, net blotch also occurs on wheat, oats, triticale and many other Gramineae. It is not a disease of major economic importance although its incidence and severity in the United Kingdom has increased significantly since the late 1970s. It is also a disease which is common in the cooler climates of north-west Europe and Canada. A spot blotch phase of *P. teres* occurs on barley in some highland areas, particularly the Andes of South America.

Disease symptoms appear first as small brown spots or blotches, often near the tip of the leaf blade. As these spots darken in colour, they elongate and develop an internal brown pigmentation in a reticulate or net-like pattern. Asexual reproduction occurs at this stage with large, cylindrical, four to six septate conidia being produced on dark, olivaceous brown, septate conidiophores which may be solitary or in groups of about two to three. The fungus is a member of the Pleosporales in the Loculoascomycetes of the Ascomycotina, the asexual stage being traditionally called *Helminthosporium teres*.

P. teres is not a high-sporing fungus, the conidia being more important if they contaminate the grain than as agents of secondary cycles of infection. Seed infection occurs when spores land on the inflorescence, germinate and then grow between the hulls and within the pericarp of the developing seed.

The sexual stage, with the production of black perithecia (more correctly, pseudo-perithecia), have been reported on overwintered crop debris. The ascospores, which are light brown, three septate and much constricted at the septa, can infect barley but they do not constitute a major inoculum hazard. In general, the disease is more important where untreated seed is sown and where stubble clearance has not been carried out effectively where barley follows barley. Infection is more severe at fairly low soil temperatures (10–16 °C) which, by slowing down the growth of the plant, extends the susceptible period. The critical stage for infection from seed-borne inoculum is before the seedlings emerge through the soil.

Control

Seed treatment with an appropriate fungicide will control the seed-borne inoculum. Traditionally, organomercurial fungicides were used but their enforced withdrawal in some countries has resulted in captan, maneb and thiram being used as alternatives, often combined with carboxin (which gives additional control of *Ustilago nuda*). Other fungicides including mixtures of gauzatine and imazalil, thiophanate-methyl and imazalil or triadimenol and fuberidazole, have been used with good effect. Control of head infection can be obtained by spraying at heading and one week later with a mixture of carbendazim and mancozeb. Cultural control measures should be practised and sources of resistance have been identified in a number of countries and are now incorporated into breeding programmes.

Southern Leaf Blight of Maize: *Cochliobolus heterostrophus*

This is one of three leaf diseases of maize in the United States that are better known by the name originally given to their asexual stages, Helminthosporium. *H. maydis*, the asexual stage of *C. heterostrophus*, is of world-wide distribution but achieved notoriety in 1970 by inducing the devastating southern leaf blight epidemic in that year. The pathogen also attacks teosinte and sorghum.

The epidemic of 1970 was caused by circumstances which have had a salutory effect on plant breeding and the teaching of epidemiology. The maize crop in the United States had become almost exclusively hybrid cultivars, developed from inbred lines whose female parents were male-sterile. Male sterility is inherited through the cytoplasm and the big mistake was to develop the most popular cultivars of that period using the so-called Texas male-sterile (*Tms*) cytoplasm. Such cultivars had a successful commercial run of about five to ten years up to 1970, during which time the T-race of *H. maydis* increased in the population. All that was needed for a severe epidemic was a conducive growing season and 1970 was characterized by persistent warm, moist weather.

The symptoms of southern leaf blight differ according to the attacking race. The common race 0 induces tan-coloured leaf lesions varying from minute spots to 3 cm long and limited by the veins. As the lesions age they become grey due to the presence of the dark golden brown, slightly curved, multi-septate conidia. With race T, the lesions are more diffuse and there may be marginal chlorosis and leaf tissue collapse. The fungus is classified in the Pleosporales of the Loculoascomycetes in the Ascomycotina.

The conidia are wind and rain-splash-dispersed, the primary source of inoculum being crop debris, especially standing stubble. Infection occurs either by direct penetration or through stomates and a host-specific toxin is produced which accounts for much of the chlorosis and necrosis. The sexual stage of *H. maydis* (*C. heterostrophus*) is characterized by beaked perithecia containing numerous asci, each of which contain filamentous ascospores arranged in coils, but it is the secondary cycles of conidia production that quickly build up the epidemic.

Control

The use of resistant hybrids in 1971 and subsequent years prevented a repeat of the epidemic, race T being only mildly virulent on genotypes with 'normal' or other *ms* cytoplasm. Crop rotation, burying or burning crop debris are useful control measures. In sweet corn, growers practise both sanitation and crop rotation and fungicidal treatments either as sprays or dusts are also effective. The dithiocarbamates are regularly used and there is an increasing use of these protectants in mixtures with systemic fungicides.

The other leaf blights of maize are caused by *H. turcicum* (*Trichometasphaeria turcica*) and *H. carbonum* (*Cochliobolus carbonum*). The former is of world-wide distribution but the latter appears to be only important in the United States. *H. turcicum* causes 'northern' leaf blight which occurs mostly in the cooler maize-producing regions and in southern Florida where the crop is grown during the winter. All three produce somewhat similar symptoms and the epidemiology and control measures are identical to those described for *H. maydis*.

Cottony Rot of Vegetables: *Sclerotinia sclerotiorum*

This is a disease which attacks a wide range of crop plants in cool-temperate and in subtropical regions. The hosts include artichoke, carrot, chrysanthemum, cucumber, lettuce, potato, tomato and many legumes. It has many common names but it is more often known simply as Sclerotinia rot. The disease causes devastating losses in the field and especially in transit and in store.

The early symptoms are the production of white, cottony mycelium over the infected parts in which black resting bodies, or sclerotia, develop later (Plate 3.1(iii)). The sclerotia are normally about the size of a pea seed but with an irregular shape.

If seedlings are attacked, a rot of the lower stem may cause a damping-off effect. Similarly, the lower leaves and stem bases of lettuce may be attacked causing the sudden collapse of the plant. With stored vegetables, especially carrot, there may be secondary spread from an infected root which can quickly cause pockets of badly rotted roots, sometimes leading to the collapse of the stored crop as secondary soft-rotting bacteria colonize the infected roots.

As infected plant parts decay, sclerotia will be released. These may survive overwinter, germinating in the spring or early summer to produce the small, saucer-shaped, stalked apothecia which represent the sexual stage of this pathogen. Sclerotial germination occurs optimally between 12 and 20 °C if sufficient moisture is available. A layer of asci covers the upper, concave surface of the apothecium, the ascospores within being expelled and either wind-blown or rain-splashed to new plant tissue. Some sclerotial germination also takes place in the autumn and apothecia may be produced, or simply vegetative hyphae will grow out which can infect neighbouring plants. The fungus belongs to the Helotiales in the Discomycetes of the Ascomycotina but there is no asexual stage of this Sclerotinia species.

Control

The eradication of sclerotia is a prime objective, soil sterilization, either by steam or by chemicals, being practised under glass. Sclerotial eradication is much more difficult in the field although, in Florida, the soil is flooded for four to six weeks in the summer to promote sclerotial decay. A rotation with rice will facilitate this approach as this crop commences its growth under flooded conditions. Small outdoor areas can be treated with drenches such as formaldehyde, the top 15 cm (6 inches) of soil then being replaced with uncontaminated soil.

All plant debris should be collected as soon as possible to minimize sclerotial production but this material should be destroyed and not composted. The viability of sclerotia is about three years but rotations are difficult due to the wide host range.

Low temperatures in transit and storage will help to minimize losses of fleshy roots. Chemicals are not yet used in control other than as soil sterilants. Resistance is also unlikely to be found as the pathogen is particularly well adapted to any carbohydrate-rich storage tissue as the host range indicates.

Other Sclerotinia diseases are also of economic importance. These range from clover rot (*Sclerotinia trifoliorum*) to brown rot of apples and pears (*Sclerotinia fructigena* in Europe) and *Sclerotinia fructicola*, the causal organism of brown rot of stone fruit.

Common Bunt or Stinking Smut of Wheat: *Tilletia caries*

Wheat is attacked by several related species of *Tilletia*, the bunt organism. *T. caries* is historically the best known but has been virtually eliminated by seed treatment. *T. foetida* causes almost identical symptoms but has a narrower distribution than *T. caries* and is not found in the United Kingdom. The two species can be differentiated on the basis of teliospore shape and by the texture of the spore walls with *T. caries* having globose, brownish black spores (15–23 μm in diameter) and with reticulate marking and *T. foetida* having globose to elongated spores (17–22 μm in diameter) but with smooth walls. *T. controversa* causes dwarf bunt and, whilst it is difficult to distinguish from *T. caries* morphologically, infected plants are very much more stunted by the former pathogen. *T. indica* is primarily a disease of bread wheat, often less severe than *T. caries* with very large, black spores (25–30 μm), more or less spherical and having a slightly rough surface. *Tilletia* is classified in the Ustilaginales of the Basidiomycotina.

Plants infected with common bunt (*T. caries*) show the first symptoms when the ears emerge. Diseased ears are bluish-green and of a very untidy appearance with glumes being spread apart due to the infected grain being larger than normal. The grain is often discoloured due to the interior being filled with the black teliospores. Breaking open the often flimsy remains of the pericarp releases the spores and a pungent, distinctly fishy smell; the spores produce trimethylamine, hence the old name 'stinking smut'.

The spores will normally be released during the combining operation but some seed may fall to the soil where they can persist. Harvested seed may become contaminated with bunt spores in the combine and initiate new disease when sown. Germinating teliospores form a promycelium the apex of which produces small projections upon which eight to sixteen basidiospores are formed, two per projection. These hyaline, filiform, slightly curved basidiospores fuse in pairs forming H-shaped structures, the fusion reconstituting the dikaryon phase which is the normal nuclear constitution of somatic cells of the Basidiomycetes. Secondary spores, or conidia, are formed on the fused basidiospores and these can germinate in moist conditions to infect the emerging coleoptile. Low soil temperature and high soil moisture are conducive to infection, the optimum being between 9 and 12 °C and around 13% soil moisture.

The mycelium of the pathogen grows within the seedling, advancing as the plant develops, eventually invading the growing point and ultimately replacing the interior of the developing grain with teliospores.

There is evidence of physiological races and new races have been produced experimentally by hybridizing two existing races (Holton, 1951).

Control

Common and dwarf bunt can easily be controlled by seed treatment. Organomercurial compounds have been traditionally used but copper carbonate dusts and formaldehyde also give good control. The use of resistant varieties would seem to be the obvious control strategy but the ease with which bunts can be controlled with chemicals has tended towards a neglect of breeding for resistance. However, it must also be pointed out that there have been reports of resistance to hexachlorobenzene in Australia.

Loose Smut of Wheat and Barley: *Ustilago nuda*

The introduction of certified seed and systemic fungicides has relegated this once most important ear disease to relative obscurity. It is a very conspicuous disease in the field, the smutted ears emerging slightly in advance of the ears of healthy plants and the rachis being covered with a mass of black teliospores, sometimes still covered by a flimsy membrane which is the remnant of the seed coat (Compendium Plate 3(iii)).

Movement of the ear by wind or implements will easily rupture this membrane, releasing the mass of spores, some of which may be impacted on the stigma or pericarp surface of neighbouring healthy plants. Germination of the teliospore produces a short promycelium after the fusion of the two nuclei. This dikaryotic state is typical of the Basidiomycotina of which *U. nuda* in the Ustilaginales is a member. The diploid, fused nucleus undergoes meiosis and the four haploid daughter nuclei migrate into the promycelium which then segments by septation into four cells. The dikaryotic state is then quickly reconstituted by means of short conjugation tubes through which one nucleus can pass into a cell containing a compatible mating-type nucleus. From this dikaryotic cell, new branching hyphae grow out and penetrate the host, often through the pericarp wall.

After penetration, the mycelium remains dormant until the seed is sown the following season. As the seed germinates and the plant grows, the mycelium develops, maintaining a position just behind the growing point until it reaches the developing ear where the mycelium completely replaces the developing floral organs to, eventually, produce the final symptom picture on ear emergence.

Physiological races have been identified but they seem to differ from country to country. The forms of the pathogens on wheat are morphologically similar although there are differences in teliospore germination. However, there is no cross-infection between the forms on the two hosts.

Control

Seed certification and its attendant inspection of the seed crops has resulted in a dramatic reduction in diseased seed. In addition, the introduction of the systemic fungicide carboxin (2,3-dihydro-5-carboxanilido-6-methyl-1,4-oxathiin) has been very successful. This chemical seed treatment is also formulated with thiram or with organomercury.

Where chemical control is inappropriate, a hot water treatment (Doling, 1965) can be used. The seed is immersed in water from 1.5–2 hours at 49 °C to 5–6 hours at 41 °C (for wheat) or 1.5 hours at 49 °C to 5 hours at 41 °C (for barley). A sliding scale of exposure time can be used which is in inverse proportion to the temperature between the above limits. Resistance genes have been identified in both wheat and barley but the convenience of chemical control makes this treatment the one normally selected.

Common Smut of Maize: *Ustilago maydis*

Common smut is a disease which has occurred wherever maize is grown although control measures appear to have eliminated the pathogen from Australia and New Zealand. Losses can amount to as much as 60% when highly susceptible sweet corn cultivars are attacked, although average losses in the United States are probably no more than 2% of the maize crop.

The extent of damage depends on the location and size of galls which are formed on infected plants. Galls can occur on all plant parts but are most common on young developing tissues or on any damaged area. Injury caused by detasselling equipment provides ample entry points, but galls are also common at stem nodes, axillary buds, individual male and female flowers, leaves and, of course, on the ears. Galls on young seedlings can seriously stunt the growth and may even cause the death of the plant. Stem galls can cause bending and a reduction in yield; infection through the silk results in the replacement of the seed by smut galls. The galls may grow to about 10 cm in diameter and are a glistening white in the early stages but then turn black as the teliospores are produced, eventually being covered only by a flimsy membrane which ruptures to release the sooty spore mass. Gall formation is caused by hypertrophy and hyperplasia of the host cells induced by the invading fungal hyphae.

The spores are wind-borne, hardy and can survive several years on crop residues. The spores are thick-walled and are formed directly from cells in the binucleate mycelium, each cell rounding off to form a spore. This method of spore production is typical of chlamydospore production in other fungi but, whereas chlamydospores are asexual spores, the smut spores are the organs of karyogamy and meiosis and the term teliospores is probably more mycologically accurate.

The teliospores germinate to produce a septate promycelium upon which ovate, hyaline sporidia (basidiospores) are produced which bud profusely in a yeast-like manner. Sporidia are also wind-dispersed and, after being deposited on a susceptible host plant, infection can occur but only if two compatible sporidia (haploid) have fused and a binucleate infection hypha is produced. However, infection hyphae can arise directly from the germinating teliospores, presumably such spores failing to undergo meiosis.

Unlike the smut fungi which attack small grain legumes, seed-borne teliospores are not the major source of inoculum as the pathogen can persist on crop debris and as a soil saprophyte. Most infections are of a localized nature on the plant although systemic colonization can arise from the infection of young seedlings. Germination of the teliospores occurs over a range of temperatures from 10 to 35 °C, the infection hyphae penetrating directly, through wounds or through stomata. There can be several cycles of infection, gall formation and spore production during a season and, in Central America where two successive maize crops are grown in the rainy season, it is easy to appreciate the potential for the build-up of inoculum. The pathogen is a member of the Ustilaginales in the Hemibasidiomycetes of the Basidiomycotina.

Control

Rotational strategies are ineffective with this disease due to the hardiness of the spores and because of the intensity of maize cropping which can provide an abundance of inoculum that can be dispersed long distances by the wind. Seed treatment can help to minimize infection from the source but the most effective control is effected by the use of resistant cultivars although the existence of physiological races of the fungus makes this more difficult. The dent types of field corn are generally more resistant than the flint types but most cultivars of sweet corn are susceptible.

Black Stem Rust of Cereals: *Puccinia graminis*

Black stem rust is distributed generally throughout the wheat growing areas of the world. The pathogen, the type species of the cereal rusts, exists in various forms adapted to particular hosts. *P. graminis* f. sp. *tritici* infects wheat, barley and triticale, *P. graminis* f. sp. *secale* infects rye. It has been responsible for very serious losses in the United States, Canada, Australia and India but is rarely of importance in the United Kingdom. It is a member of the Uredinales in the Basidiomycotina.

Disease symptoms occur commonly on stems and leaf sheaths but leaf blades and ears may also be attacked. The pathogen is **heteroecious**, alternating between the cereal host and the secondary host, the common barberry (*Berberis vulgaris*). Three spore forms are produced on the cereal host. The first can be seen in elongated, brown urediosori which erupt through the epidermis of leaves and stems. The **urediospores** produced in these sori are ellipsoid or ovoid, golden-brown, $21-42 \times 16-22\,\mu$m, echinulate and with four equatorial germ-pores. They continue to be produced until the plants approach maturity. As the season progresses, the urediosori are replaced by black teleutosori which often merge to form black, powdery streaks, especially down the stems. The **teleutospores** are dark brown, two-celled, stalked and somewhat wedge-shaped. They measure $35-60 \times 12-22\,\mu$m. The teleutospores can persist in plant tissue overwinter. On germinating, they produce an out-growth, the basidium, which divides up into four cells, each cell having a single haploid nucleus, the four products of a meiotic division which occurred when the two nuclei in a cell of the teleutospore (the normal dikaryotic state) fuse to give a diploid nucleus. Each cell of the basidium produces a single **basidiospore**, or sporidium; these spores cannot infect wheat but are wind-dispersed and can only infect the **alternate** host, the barberry.

On the barberry, the pathogen produces flask-shaped pycnia on the upper surface in which masses of minute uninucleate, **pycniospores** are produced which, eventually, are extruded through the ostiole in a sugary carrier substance which attracts insects which then passively disperse the spores which stick to their body surface.

Puccinia graminis is a **heterothallic** fungus with both mating types being represented in the four basidiospores. The pycniospores will, consequently, be of the same mating type as the basidiospore which produced the infection. The sexual stage begins with the fusing of a pycniospore of one mating type and flexuous hyphae which emerge from pycnia of the opposite mating type. In fusing, the recipient cell is reconstituted as a dikaryon and further colonization of the leaf occurs with the production of cup-shaped aecidia on the lower surface. Masses of **aecidiospores** pack the base of the aecidium as polyhedral shapes (Compendium Plate 4(iii)). As the pressure is released towards the mouth of the aecidium, the spores assume a more global appearance. The aecidiospores are dikaryotic and wind-dispersed but can only infect the cereal host.

Infection of wheat by aecidiospores or the clouds of urediospores that can rapidly cause severe epidemics, occurs through the stomates. A film of moisture is required on the host surface for infection to take place and the disease develops most rapidly in wet, cool weather (18–20 °C).

At least 200 physiological races of *P. graminis tritici* have been identified and these can curtail the useful commercial life of cultivars if race-specific resistance is employed.

Control

In North America, the eradication of barberry and the use of resistant cultivars have significantly reduced the impact of this disease. In the United Kingdom, the pathogen does not overwinter on barberry, the inoculum for the very rare epidemics being blown up from the Iberian peninsula. Resistance has been the main strategic measure against stem rust with a recent change of emphasis towards the incorporation of race non-specific resistance or more durable race-specific resistances. Diversification would help to reduce disease levels but, in recent years, there has been a considerable increase in the use of systemic fungicides, in particular dichlobutrazol and a mixture of carbendazim and dithiocarbamates with carboxamides, morpholines and the ergosterol biosynthesis inhibitors also being successful.

Other Rust Fungi

The Uredinales comprises very important plant pathogens. In the genus *Puccinia*, *P. coronata* causes crown rust of oats and has proved very damaging in North America and Australia although it has declined in the United Kingdom with the reduction in the growing of the oat crop. It is another heteroecious rust with the buckthorn (*Rhamnus cartharticus*) as the alternate host. The urediospores are orange and the black teleutospores are characterized by the crown-like appendages at their apex. *P. hordei* causes brown rust of barley, the urediospores being yellowish-brown and being produced in scattered urediosori on the leaves, the sori often having a chlorotic halo surrounding them. The sexual stage is rare and insignificant epidemiologically. *P. polysora* causes southern corn rust and attacks *Tripsacum* spp., *Erianthus* spp. and *Euchlaena mexicana* as well as the *Zea mays* crop. It is now one of the most important diseases of maize in West Africa and can be particularly severe in some areas of the United States, North Carolina for example. Rust pustules appear on all plant parts but most commonly on both leaf surfaces. There is no known alternate host. *P. recondita* causes brown rust of wheat and rye and it can be very damaging on the former host where it often causes more damage than *P. graminis tritici*. There are three known alternate hosts, the most common being meadow rue (*Thalictrum* spp.). *P. sorghi* is a universal rust on the maize crop. It is generally unimportant although severe infections can cause yield reductions in sweet corn. The urediospores are cinnamon-brown and the chestnut-brown teleutospores are rather oblong in shape and constricted at the septum. Overwintering is mainly as teleutospores on crop debris, basidiospores being produced in the spring which can infect several species of *Oxalis*, the alternate host upon which aecidia are produced. *P. striiformis* causes stripe or yellow rust of wheat and barley and is particularly important in north-west Europe and the United Kingdom. The urediospores are lemon-yellow and the urediosori are produced in lines between the veins to give a stripe effect. No sexual stage is known and overwintering depends on the presence of the cereal host.

Phragmidium spp. cause rust diseases on a number of hosts including rose and blackberry. They can easily be distinguished at the teleutospore stage as these have long stalks and are multi-septate. *Hemilea vastatrix* is the causal organism of coffee rust. It attacks leaves, petioles and growing twigs, bean-shaped, orange urediospores being pro-duced just below the epidermis. When the epidermis ruptures, the spores are exposed and can be dispersed by rain-splash and wind. Teleutospores are seldom produced. Long distances have been quoted for the dispersal of this pathogen and there is some evidence that the disease entered Brazil in 1970 after the spores had been blown across the Atlantic Ocean although many pathologists are inclined to the view that the disease entered on infected plants. The most effective way to control coffee rust is to use resistant cultivars although spraying with copper or dithiocarbamates, and some systemic compounds, can keep the disease in check.

Grey Mould of Lettuce: *Botrytis cinerea*

This is an almost ubiquitous fungus of great importance under certain environmental conditions and attacking a wide range of plants including cucumbers, broad beans and tomatoes. It can be particularly damaging on lettuce in cool, wet seasons. In the seedling stage, it resembles the common damping-off diseases (*Pythium* and *Phytophthora* spp.). On more mature plants, a soft brown rot may occur on the stem base and lower leaf petioles eventually progressing to the heart of the lettuce causing a slimy disintegration of the tissue.

The leaves become limp and discoloured, often having a reddish colour at the base, with a grey, superficial mould appearing on the surface, particularly on the underside. Microscopic examination reveals masses of dichotomously branching conidiophores emerg-ing from stomata bearing several conidia on each branch tip. The disease can progress through the head until all the leaves are transformed into a slimy mass. It is also possible to pick up the wilting lettuce head from the soil with little or no anchorage resistance, the inner rot having disintegrated the lower stem base.

When the rot is well advanced, flat or cylindrical mycelial resting bodies, the sclerotia, are formed and will be released into the soil upon complete decay of the host tissues. These overwintering bodies normally germinate the following season to produce asexual conidiophores and conidia which can initiate new infections.

The conidia germinate quickly in the presence of water and the germ-tubes enter through wounds, often caused by *Bremia lactucae* infections. Spores are dispersed by both air and water-splash and the disease is favoured by humid conditions, temperature being less critical, the pathogen being active over a wide range up to 25 °C. *B. cinerea* is the asexual stage of *Sclerotinia fuckeliana* but the sexual stage is not found under field conditions although it has been produced in laboratory culture. It is a member of the Discomycetes of the Ascomycotina.

Control

All diseased plants and plant debris should be carefully removed and destroyed. Damage or any conditions leading to senescent tissue should be avoided. Under glass, overwatering should clearly be avoided. Any techniques which reduce humidity, such as heating the air or ventilation should be practised. Protective treatments are advised from the seedling stage onwards. Seed beds may be dusted with quintozene before sowing or dicloran which can also be applied at intervals of about six weeks. It is important with all foliar-applied fungicides to protect the lower leaves before subsequent growth prevents the spray reaching them. Routine sprays with benomyl, carbendazim or thiophanate methyl after planting out will minimize losses, and high volume sprays of iprodione can also give good control.

Seedling Blight, Foot and Root Rot and Head Blight of Cereals: *Fusarium culmorum*

This disease causes a variety of symptoms on many cereals and grasses and can cause serious losses, especially in wheat, with reductions of up to 50% being reported from many countries. *F. culmorum* is the only one of four species of *Fusarium* attacking cereals in which the sexual stage is unknown.

The fungus is a highly competitive 'soil-inhabiting' organism which is very resistant to cold temperatures and which can survive for two years on buried stubble trash. When young seedlings are attacked, they may be killed even before emergence. The symptoms on surviving seedlings are brown streaks at the base of the coleoptile but these symptoms are almost identical to those caused by other *Fusarium* spp., *F. avanaceum* and *F. nivale* for example.

Infection of the coleoptile is favoured by any condition which retards seedling emergence, poor drainage and acid heavy soils having this effect. Plants which survive the pre-emergence attack may later succumb to foot and root rots especially if there is a humid microclimate within the crop. This 'brown root rot' stage is characterized by a brown discoloration of the basal leaf sheaths often followed by a more pronounced rotting of the tiller bases at about soil level. Severely affected plants may either fail to produce heads or develop the classical 'whiteheads' with empty ears. Infection of the ears can also be recognized by the production of coral-coloured conidial pustules, **sporodochia**. There may also be a brown spotting on the glumes and the whole ear may become tinted a red/pink colour. Infection of ears affects grain yield depending on the length of time infected, an early infection producing the most shrivelled grains or the most empty heads.

The pathogen is classified as a member of the Hyphomycetes in the Deuteromycotina although many other *Fusarium* spp. are classified in the Ascomycotina on the basis of their known sexual stage.

Control

The disease can be controlled by organomercury seed treatment but this is prohibited in many countries. Alternatives are carbendazim, fuberidazole, thiabendazole and thiophanate-methyl. In addition, many currently recommended foliar sprays to control other cereal diseases are also active against *F. culmorum*. The burial of crop residues will help but, in the long term, control will probably be best achieved by the development of resistant cultivars.

Verticillium Wilt of Tomatoes: *Verticillium albo-atrum* and *V. dahliae*

Verticillium wilt, or 'sleepy disease' may be confused with another wilt disease caused by *Fusarium oxysporum* f. sp. *lycopersici* or simply with a lack of water. The lower leaves begin to show chlorosis, often on one side only, while the upper leaves curl upwards and inwards. Slicing through the stems, petioles or roots with a knife will reveal a characteristic brown discoloration in the vascular tissue. The leaves droop in the daytime and, although they may make a recovery during the night, the amount of wilting increases daily although it may take several weeks to kill the plant.

Verticillium is a soil-borne fungal pathogen which, although not now very common, can be a devastating disease. It has a wide host range, affecting cucumbers, chrysanthemums, lucerne, peppers and Brussels sprouts, and although isolates from one host may infect many others they never produce the level of damage as they do on the host species from which they originated.

Entry is through the roots, either by direct penetration or after damage, the central vascular system eventually being colonized and blocked by mycelial growth. The pathogen produces pectic enzymes which degrade cell wall components and, if tomato fruit are invaded, a soft rot results. Toxins may also be produced in the latter stages of disease development and can cause cell death.

In the absence of growing plants, the fungus survives on dead roots and other infected debris and the hyphae from such inoculum sources can grow out and infect the roots of the subsequent crop. Asexual reproduction can also occur on infected debris under moist conditions. Aerial conidiophores develop upon which several whorls of short branches bear single conidia at their tips. Great numbers of conidia are produced but experiments have indicated that without large numbers there would be little effect on yield. Small, dark resting bodies, or sclerotia, are also produced in infected plant tissue and can prove inoculum hazards as they can survive for long periods. There is no known sexual stage and *Verticillium* is classified in the Hyphales of Deuteromycotina.

Disease development is closely related to temperature, *V. albo-atrum* having a growth optimum of 20–25 °C and *V. dahliae* 25–28 °C.

Control

Under glasshouse conditions, raising the temperature above 25 °C (77 °F) will reduce disease but only if *V. albo-atrum* is the causal organism. Premature planting in the spring should be avoided as the soil temperatures are likely to be low, predisposing the plants to infection. Host plant resistance genes have been incorporated into many commercial cultivars and such resistance is also utilized in rootstocks on to which scions of the desired cultivar may be grafted.

If susceptible cultivars are to be grown, the soil should be sterilized to the depth of root growth. Soil sterilizations by steam or a chemical fungicide is highly recommended and drenching the plants with benzimidazole fungicide can help to minimize the effects of disease although isolates of *V. dahliae* 'resistant' to high levels of this chemical have been reported. However, in general, once the disease is apparent in the glasshouse there is little that can be done, although raising the temperature can alleviate the most damaging effects.

Eyespot of Wheat and Barley: *Pseudocercosporella herpotrichoides*

Eyespot is generally assumed to be a disease of winter wheat and barley but spring cereals in many regions of the world are also often attacked, although oats and rye have a high level of resistance. The symptoms are mainly confined to the basal portion of the plant with characteristic elliptical or eyespot lesions developing on the lower leaf sheaths at or just above ground level (Plate 3.1(ii)). Young lesions have brown, elongated, diffuse borders with straw-coloured centres at first but which darken to give a 'pupil' appearance with age. A section through the stem will often reveal grey mycelium in the stem cavity behind the lesion. There may be several lesions on a tiller and eventually the whole stem-base may be circled. Severely attacked stems may break at the site of the lesions causing 'lodging' of the crop to produce a tangled mass of fallen plants, a symptom described as 'straggling'.

Infection normally occurs through the stomata and the hyphae progressively penetrate inwards. The early infection of plants can result in death; later infection and the destruction of the vascular tissue will result in partially empty heads, the 'whiteheads' symptom.

Sporulation mostly occurs during cool, moist weather in the autumn and spring and ceases as summer approaches, spore dispersal being effected mainly by rain-splash. Unbranched, short, erect conidiophores are produced on the eyespot lesions on the hyphae at the end of which, long, thin and very slightly curved conidia, normally three to seven septate, are produced. The pathogen is a member of the Hyphales in the Deuteromycotina and there is no known sexual stage.

Conidial production in pure culture only occurs under fluctuating temperatures between 0–13 °C, reflecting conditions during sporulation in the autumn and spring. Severe infection can occur with low soil temperatures of 6–10 °C, and above 15 °C, there is very little disease.

The fungus can survive the intercrop period on crop residues but volunteer wheat plants have also been shown to act as secondary hosts. Heavy applications of nitrogen increase disease severity and, by weakening the straw, also exacerbate lodging. The following factors, either singly or in combination, can contribute to the possibility of high levels of disease occurring: intensive cereal cultivation; high nitrogen applications; early drilling of crops; heavy soil; mild, wet winter; susceptible cultivars.

In the field, eyespot symptoms can be confused with those caused by *Rhizoctonia cerealis*, the sharp eyespot pathogen. The presence of mycelium in the stem cavity of the former and the sharply defined lesion margin and the pale or cream colour of its centre in the latter, are reasonably reliable diagnostic features.

Control

Cultural practices aimed at reducing the inoculum by the destruction of volunteers and infected debris and by rotation, should form the basis for other management strategies. Many cultivars are now available with high levels of resistance and, where possible, cultivars with short, stiff straw should be grown. The application of straw-shortening chemicals such as CCC (chloride-choline-chloride) can also be of benefit.

The application of fungicides for eyespot control has long been recommended for high risk situations. Systemic chemicals can be used either as seed-treatments or as early foliar sprays. Unfortunately, there has recently been a large increase in resistance of *P. herpotrichoides* to MBC fungicides and these chemicals should only be used in mixtures. Fungicide application is advised for high-risk areas as soon as the first node is detectable or, if eyespot is obvious and penetrating through two or more leaf sheaths, on at least 20% of tillers at the leaf sheath erect stage. Mixtures of prochloraz and MBC fungicide are widely recommended and these can also be used in low-risk areas either as a routine, insurance spray or as part of a programme to control other pathogens such as mildew and Septoria.

Leaf Blotch or Scald of Barley: *Rhynchosporium secalis*

This is an ubiquitous disease in all the barley-growing regions of the world. In the cooler maritime regions it can be particularly damaging with large reductions in leaf photosynthetic tissue with yield reductions of around 20–30% often being reported when susceptible cultivars are grown.

The symptoms on the leaves (Compendium Plate 2(iv)) are easily recognizable, large (1 cm or more), irregular or diamond-shaped blotches with dark brown or purple edges scattered over the surface, often on the leaf margins and, with a high level of infection, coalescing to produce extensive patterned diseased areas. Stems and awns may also be attacked.

The centre of the lesions have a water-soaked, bluish-grey appearance in the early stages when abundant conidia will be produced. As the lesion ages, the centre may become gingery-brown and paper-thin, often tearing and giving the diseased leaf a very tattered appearance. The conidia are borne on very short, rudimentary conidiophores which are merely raised areas of the hyphal cell wall. The conidia are two-celled, cylindrical to ovate with a characteristic apical beak. These asexual spores are dispersed by rain-splash and will germinate in the presence of a film of moisture on the leaves. The disease is favoured by cool, wet weather with the optimum temperature for infection 15–18 °C although infection can occur between 12 and 24 °C.

The pathogen can infect seed but the bulk of the overwintering inoculum is derived from infected crop debris, from the autumn-sown crop and from infected volunteer barley plants. Rhynchosporium can also attack certain grasses which can act as reservoirs of inoculum. The fungus is a member of the Hyphales in the Hyphomycetes of the Deuteromycotina and, as such, has no sexual stage.

Control

Management strategies are mainly concerned with the use of resistant cultivars or the application of fungicides. Both race-specific and race non-specific resistance are used, major genes numbering 1–11 having been recorded. The fungus exists as physiologically specialized forms which does introduce problems as regards the commercial life of race-specific resistant cultivars. Fungicides should be applied in the autumn to winter barley if about 10% of the lower leaves is affected. In the spring, if the disease can be seen on the youngest leaves, another application should be made, often with the aim of controlling other pathogens as well. At ear emergence, if the disease is present on any of the top three leaves of autumn or spring sown barley, a final application should be made. Chemicals commonly used include benomyl and other MBC fungicides, prochloraz, propiconazole, pyrazophos, tridemorph and triadimenol. Some of these are also formulated as fungicide mixtures with dithiocarbamates.

Tomato Leaf Mould: *Fulvia fulva*

This is a very common and troublesome disease of the glasshouse tomato crop and occasionally in outdoor tomatoes during wet, moderately cool seasons in warm countries and areas such as the southern states of the United States.

The onset of disease can be quite sudden, being seen most often on the leaves although the stems and blossoms can also be attacked. The fruits are rarely affected. The first symptoms are light-green or yellowish spots on the upper surfaces of the leaves. Corresponding with these spots on the lower surface are areas covered with an olivaceous or pale brown, velvety mould which is the asexual stage of the fungus comprising conidiophores and conidia. As colonization of the leaf tissue increases, the attacked areas become brown in colour and, eventually, the whole leaf dies.

The conidia are readily spread by draughts and air currents to infect other plants and rapidly cause a serious outbreak. The conidia require a humid atmosphere for germination when they produce a germ-tube which infects the host tissue through the stomata. The conidia are very resistant to dry conditions and to low temperatures and can survive the intercrop period. Mycelial fragments can also survive in infected plant debris. There is no sexual stage of this fungus which is a member of the Hyphales in the Deuteromycotina.

Tomato leaf mould is greatly influenced by the environment, especially being favoured by temperatures above 20 °C and humidity above 95%. Unfortunately, these conditions occur all too often during the summer months especially where insufficient care and attention is given to ventilation.

Control

Cultural methods of control can do much to alleviate this disease problem. Crop debris should be removed and destroyed between crops and the spraying of the glasshouse interior with 1 part of 40% formaldehyde in 49 parts of water is recommended. The burning of sulphur can also be practised to disinfect the glasshouse but only with wooden structures as corrosion is a problem with aluminium houses.

Adequate ventilation to prevent stagnation below the crop canopy is critical. Watering should be carried out just prior to ventilation on hot days in order to minimize the rise in humidity and to keep them relatively dry in the night. If the disease does come in, it is advisable to destroy all the leaves below the fourth truss immediately after the fruit from the truss has been picked.

Resistant varieties offer a cheap and practical method of control although the pathogen has several physiological races and, in the United Kingdom, there are no cultivars which are resistant to all the races.

Fungicidal treatment can give good control although many of the fungicides, being protectants, require repeated application. The dithiocarbamates, zineb and maneb, give good control as does the systemic, benomyl, and the sulphamide dichlofluanid. However, chemical control measures should only be contemplated when all other management measures fail and, even then, the cost-effectiveness should be closely scrutinized.

Leaf Spot of Celery: *Septoria apiicola*

This most damaging disease of celery was once thought to be caused by two distinct species of *Septoria*. Only *S. apiicola* is now recognized and is classified in the Sphaeropsidales of the Deuteromycotina even though its pycnidia and pycnidiospores are very similar to those of *S. tritici* (*Mycosphaerella graminicola*, see *Leptosphaeria nodorum*). No sexual stage of *S. apiicola* has been identified.

The symptoms are best seen on the older leaves where groups of black pycnidia can be seen on spots which ultimately turn brown and which are surrounded by a distinct chlorotic halo. The spots are usually very numerous and eventually coalesce so that the leaf shrivels and collapses. This is the 'blight' stage, the disease sometimes being referred to as celery late blight. Pycnidia may also be seen on the petioles and on the surface of infected seed. Under moist conditions, pycnidiospores which are hyaline, filiform with one to five septa and measure 22–56 × 2–2.5 μm, are extruded from the pycnidia in a gelatinous, tendril-like cirrhus where they are exposed to dispersal by rain-splash. The quality of the celery crop can be greatly reduced by the rusty brown lesions which develop on the lead stalks.

Infected seeds are the main source of inoculum but the pathogen can remain viable on plant debris for up to two years. There is also evidence that planting machinery can also spread infected crop debris from site to site and great care should be taken to destroy infected leaves which have been trimmed off prior to packaging. An interesting facet of epidemiology is that, on germination, the seed coat is often pushed above the ground exposing any pycnidia and allowing spores to be dispersed to the emerging cotyledons and ultimately to all the foliage. Wet weather is conducive to the development of epidemics especially if temperatures are cool and slow down plant growth. Periods of dry weather which create a soil moisture deficit can also slow down the plants' development and, if these are interspersed with wet periods, disease will spread rapidly.

Control

The use of disease-free seed is essential and seed should only be purchased if seed tests confirm the absence of the pathogen. The practice of thiram-soaking is recommended for all seed and gives good control. The treatment involves immersing the seed, usually held in muslin bags, in a 0.2% (w/v) active ingredient at 30 °C for 24 hours. After this treatment, the seed should be dried under forced air at 20–25 °C for about 5–6 hours. A reasonable rotation is advisable with at least two years between successive crops. Great attention should be given to crop hygiene. Resistance is not yet available in commercial cultivars.

Diseases Caused by Bacteria

Crown Gall: *Agrobacterium tumefaciens*

This is a very successful soil inhabitant which causes the deformation of a very wide range of plant species. Apples, pears, cherries, plums, cane fruits, grapevine, tomatoes and many herbaceous and perennial ornamentals are commonly attacked producing quality reducing disfigurations. Hops appear to be the only crop which can be killed by this bacterium.

It is a classical wound parasite, infection being made possible by mechanical damage or by insect punctures. Grafting operations afford easy entry of this pathogen. The host cells invaded are stimulated into a cancer-like phase of division to produce pronounced and often unslightly galls. The mechanism of this hyperplastic activity has only recently been elucidated. On entering the host cell, the bacterium injects a small, circular piece of DNA which has been named the *Ti* (tumour-inducing) plasmid. This phenomenon has attracted much interest from molecular geneticists and plant breeders who recognize a potential for introducing new genetic material into eukaryotes. The genetic information, on being integrated with the host's chromosomes, codes the cells not only to divide but to produce a series of compounds called opines which are used uniquely by the pathogen. The galls produced vary in size and texture according to the host infected. Most are more or less round, distinctly lobed and varying in size from a centimetre or so in diameter to large (Plate 3.2(ii)), hard gnarled swellings, often about 20 kg in weight, on larger deciduous tree hosts. Mostly, as in the case of chrysanthemums and roses, the disease is relatively harmless although, with the disfigurement, quality and hence crop value may be significantly reduced. There is also evidence that some scion cultivars of cherry predisposed the root-stocks to infection.

Control

The long-term approach to control is through the production of clean nursery stock which can only be effected by rigorous inspection and certification. Care must also be taken to minimize all forms of wounding and grafts should be protected from infection either by the application of grafting wax or by wrapping the union in adhesive tape. There are differences in susceptibility between commonly used rootstocks of apple, cherry and cultivated blackberry. Biological control procedures have been developed in Australia to control the disease on stone fruits. The technique involves the use of a pre-plant dip in a suspension of *Agrobacterium radiobacter* strain 84 which, by producing a bacteriocin, is antagonistic to some strains of the crown gall bacterium. Although the bacterium has been shown to be able to survive in soil for up to two years, normal soil fumigation has not proved to be successful.

Fire-blight of Apples and Pears: *Erwinia amylovora*

This disease has a unique place in history being the first one, in 1878, in which a bacterium was shown to be the causal agent. The disease is notoriously unpredictable and can cause extensive damage to pears but is somewhat less serious on apples. The pathogen is apparently indigenous to North America, but the disease is also important in Italy, Japan and New Zealand. It was first reported in the United Kingdom in 1957 and caused such severe losses in the next few years that the disease was declared notifiable and strict control measures enforced.

The first symptoms appear on the blossoms, sepals, leaves and twigs, the affected blossoms and sometimes shoots wilt and turn dark brown or black. Leaves may be affected at the margins or on the blade with the lesions often covering the whole blade. The bacteria quickly spread through the tissues of the fruiting spur as summer progresses and eventually can kill the branches. The death of twigs and branches and the earlier death of blossoms and leaves which remain on the trees, give the infected area a very scorched appearance, hence the common name (Plate 3.2(iv)). There may be several such infected areas on a tree, each being termed a 'strike'. Infected blossoms can be confused with the blossom-blight caused by *Pseudomonas syringae* and only laboratory tests will positively confirm the identity. When the bacterium reaches the main trunk, it can quickly invade other branches and the tree will usually die soon afterwards.

Cankers can readily be seen on young bark as dark, shallow, often water-soaked areas with an indistinct border between healthy and infected tissue. In humid weather, amber-coloured droplets of bacterial ooze will be exuded over the recently invaded parts, the onset of dry weather causing the droplets to dry up to give a silvery film over the bark. The cankers become inactive during the winter, they are often called 'hold-over' cankers, and their outline is more pronounced because of cracks which appear at the margins. Fruits formed on infected spurs become very wrinkled, almost black and with a distinctly oily appearance. Symptoms on apples are very similar but never as extensive.

The risk of fire-blight is greatest above 18 °C when there is rain. Blossom infection is particularly likely on sunny, warm days (21–30 °C) when insect activity is high. Rain has been shown to be a major factor in the dissemination of the pathogen although spread is also effected by pollinating insects and by the wind dispersal of particles of dried slime.

The 'hold-over' cankers produce abundant inoculum as the temperatures rise in the early summer, infection taking place through natural openings in the receptacle, through stomates in the sepals and leaves and through wounds. The disease spreads in the outer tissues of branches in the intercellular spaces and a thin slice of bark removed from the leading edge of a canker will reveal a reddish-brown discoloration with a distinctly wet appearance which may also be sticky to the touch. In the United Kingdom, spread during the spring has been relatively unimportant due to the low temperatures. However, cultivars with a secondary summer blossom flush are likely to be at a high risk. The pear cultivar Laxton's superb produces far more summer blossoms than other commercial cultivars, and this accounted for its high susceptibility in the United Kingdom during the late 1950s and early 1960s. This cultivar was then prohibited by legislation in certain areas of the United Kingdom.

The pathogen can infect several other roseaceous species including hawthorn, cotoneaster and pyracantha.

Control

Regular inspection is the key to good control in the United Kingdom. Both orchard crops and alternative host species should be watched and advice sought immediately symptoms are observed. In hotter countries, fungicidal sprays during the blossom period give partial control, and antibiotics like streptomycin are also fairly effective although not permitted in the United Kingdom. Winter pruning of infected twigs should always be practised, the excised material being burned immediately. It is probably advisable to grub out the tree completely if there are more than three 'strikes'. Many countries also practise severe cutting back of hawthorn hedges, and beehives should be removed from the orchard after a good set has been achieved. Every attempt should also be made to remove summer blossoms before they open.

Blackleg of Potatoes: *Erwinia carotovora* subsp. *atroseptica*

Although very common, this bacterial disease rarely causes serious losses and then only in wet seasons. However, it can cause considerable deterioration in bulk stores especially if a wet crop has been harvested. The bacterium is closely related to *E. carotovora* subsp. *carotovora* which causes a soft rot of potatoes and many other root crops, particularly if wounding has occurred.

The first symptoms are generally seen in early summer, plant death producing a 'gappy crop' giving the first indication although this would be hard to positively confirm as blackleg. The first obvious symptoms are stunted, more or less erect plants with yellowish leaves that are often stiff to the touch and rolled inwards from the margin. A more identifiable symptom is revealed if such plants are gently pulled for they normally break at soil level where the base of the stem, a couple of centimetres or two above and below the soil, is blackened and rotting but still reasonably firm.

A knife section cut vertically through an infected stem shows a brown discoloration in the woody tissue. However, not all of the stems of an infected plant may be affected. Early infections often cause the whole of the haulm to rot away. Later infections, especially in wet seasons cause rapid wilting, a more extensive soft and wet black rot of the stem base and the quick death of the plant.

The causal bacterium is common in potato-growing soils although it does not survive long in the absence of potatoes. It is quite common, therefore, for most tubers to be contaminated with cells of *E. carotovora*. If these tubers are stored under incorrect conditions they will quickly rot. Some seed tubers may be planted which the bacteria have infected but to such a slight degree that their diseased state goes undetected. There is little evidence that the disease spreads from plant to plant although the bacteria have been detected over a metre distant from an infected plant.

Control

In large bulk stores, chemicals such as dichlorophen and organoiodine complex may be sprayed on to the tubers. There is no control for the disease in the field and, whilst roguing is often carried out, it does not eliminate the bacterium from stocks but it may reduce the amount of infection carried. The planting of healthy uncontaminated tubers is emphasized but with the difficulties of recognition, it is advisable to purchase seed tubers from a reputable merchant. The practice of cutting seed tubers prior to planting should also be avoided as the bacterium is easily spread from tuber to tuber on the knife surface.

Bacterial Canker of Tomato: *Corynebacterium michiganense*

Bacterial canker of tomatoes has caused serious losses in North America, Europe and Australasia although its importance has been much reduced due to the use of clean seed and seed treatments. It is not a common disease in mainland Britain although both glasshouse and field-grown tomatoes have been badly affected in the Channel Islands.

The disease can be categorized with the vascular and parenchymatous bacterial diseases and is a rare member of the Gram-positive pathogenic bacteria and the only one likely to be found in large amounts in tomato tissue. The first symptoms are white mealy spots on the upper leaf surface which later become necrotic and which may also be associated with some wilting and marginal curling although there is considerable variation in these symptoms depending upon cultivar, time of infection and cultural technique. In an advanced stage, the diseased plants look decidedly bedraggled with necrotic, wilting and battered foliage. Internally, a brown discoloration may be found in the vascular tissue and, under certain conditions, there is a spotting and streaking of the stems which may eventually crack open. These scab-like spots or streaks on the stem are not typical of classical bark cankers but this is the stage which gives the disease its name. Affected fruits show raised, light-coloured lesions with a brown centre, often referred to as 'birdseye' spots.

Infection can take place through the roots if plants are planted into contaminated soil where the bacteria can persist for two to three years but, more importantly, damage caused during cultural operations such as trimming, deleafing and tying provide ample entry points. Seeds may become contaminated and provide inoculum for the next crop although infected seedlings from this source do not become apparent until some time after planting. High humidity is necessary for the superficial symptoms to develop and spraying with pesticides will aggravate any humid conditions.

Control

With symptom expression varying considerably, early diagnosis of bacterial canker is most difficult. Seed is undoubtedly the most important primary source of inoculum and only seed that has been extracted by the standard hydrochloric or acetic acid extraction method, or by fermentative extraction, should be used although these treatments are less successful in reducing deep-seated seed infection.

The practice of syringing with water as an aid to pollination and the application of pesticides should cease if the disease is even suspected as being present. Infected plants should always be removed and destroyed and copper-based sprays applied every three days give good protection. In addition, operator hygiene is crucial as the bacteria can be easily spread on hands, implements and overalls especially under damp conditions.

Bacterial Ring Rot of Potatoes: *Corynebacterium sepedonicum*

This is a very infectious wilt of potatoes which, by 1940, had been found in thirty-seven states in the United States. It is not known to have occurred in the United Kingdom but it is present in most of the northern European countries.

Under cool spring and summer conditions, symptoms on the foliage are negligible but, when spring is cool and summer is warm, the symptoms are a yellowing and marginal necrosis of the lower leaves on one or more stems when the plant is nearly full-grown. There may also be some stunting and the stem may collapse. Sections through the stem will show vascular browning and a creamy bacterial ooze will exude from the cut ends if the stem is squeezed.

Tubers may become infected by way of the stolons, the early symptoms being a light yellow vascular discoloration which is associated with a cheesy rot. This rot intensifies to give a dark ring in the vascular tissue which eventually breaks down the tissue structure to form cavities.

The organism survives the intercrop period in tubers although it has also been detected as dried ooze particles on crates and machinery. There is no evidence that it persists in the soil although volunteer potato plants could act as carriers between crops. Spread is mainly between tubers during mechanical handling, especially on elevators and during planting, and the practice of cutting seed potatoes is not to be recommended unless stringent aseptic routines are adopted between each cut.

Control

Legislative measures have been enforced in the United Kingdom to prohibit the entry of potatoes from countries in which the disease is endemic, as the British climate would certainly be conducive if this pathogen gained entry. In the United States, seed-certification and strict hygiene routines if contamination occurs are practised to good effect.

Yellow Slime or 'Tundu' Disease of Wheat: *Corynebacterium tritici*

This bacterial disease is of greatest importance in the Middle and Far East with severe outbreaks having been reported in China, Egypt, India and Iran. The symptoms are easily identifiable with a yellow bacterial slime covering all or part of the ear. On badly affected plants, the ear may emerge distorted or fail to emerge at all. There may also be a failure to produce grains. The symptoms on leaves and sheaths are again characteristic with long yellow spots with a distinctly slimy appearance.

Unusually for bacterial plant pathogens, *C. tritici* has a nematode vector, the 'ear cockle' nematode, *Anguina tritici*, the bacterial cells either contaminating the surface of or being present within galls of the nematode upon which they can remain viable for at least five years.

Control

Immersing the seed in 20% brine enables the galls to be skimmed off although this treatment is not totally successful as the heavier galls and those that quickly become waterlogged sink to the bottom with the seed. If carried out precisely, the hot water treatment as described for *U. nuda* can give reasonable control.

Bacterial Wilt of Maize: *Erwinia stewartii*

Bacterial wilt, or Stewart's disease, of maize occurs mainly in North and Central America but outbreaks have been reported in the Far East, South Africa and Europe. The bacterium blocks up the vascular system and the symptoms reflect this physiological upset. Long, pale green or yellowish streaks develop on the leaves, mostly starting at a point of insect damage as the main vector of this pathogen is the corn flea beetle (*Chaetocnema pulicaria*). Other vectors include the soil-borne larvae of *Diabrotica longicornis* and *Phorbia cilicrura,* both of which attack the roots. The plants wilt typically, often followed by death, but those not killed are stunted and tassels may be produced early with the ears small or even missing.

Control

The use of resistant cultivars still offers the most effective method of control with resistance being controlled by two major genes in sweet corn. In addition, reasonable control can be achieved by the use of early season application of insecticides, timed to control the flea beetle population.

Halo Blight of Oats: *Pseudomonas coronofaciens*

This is a seed-borne disease which is prevalent in many countries especially Scotland, Canada, New Zealand and the United States where it was first described. The bacterium is a motile rod with one to several polar flagella. It produces white colonies on nutrient agar, is Gram-negative and has an optimum temperature for growth of 22–25 °C.

Infected seedlings develop symptoms from the first leaf onwards, producing small brown, necrotic spots surrounded by a light green oval 'halo' which have an overall diameter of about 4–5 mm. The lesions are often found on leaf margins but leaf sheaths and glumes later become infected as the bacteria are rain-splashed over the developing plant.

Entry is effected through natural openings or injuries and insects are undoubtedly causal factors in this respect. The bacteria abound in the central dead tissue of lesions which have a 'water-soaked' appearance. The chlorotic halo is produced as a result of the cessation of chlorophyll production, possibly due to the action of a bacterial toxin. Photosynthesis is thus reduced and glume infection is usually accompanied by the prevention of spikelet development. The disease is favoured by a cold, wet spring and is absent in warm dry summers.

Control

The main aim is to ensure the sowing of healthy seed by using resistant varieties or by minimizing any infection of the seed by chemical treatments such as formaldehyde or streptomycin. Good resistance has been found in a hexaploid oat, believed to be a natural cross between *Avena sativa* and *A. ludoviciana*, but the decreasing area sown to oats in recent years had led to a neglect of resistance breeding programmes against this pathogen.

Halo Blight of Dwarf and Runner Beans: *Pseudomonas medicaginis* subsp. *phaseolicola*

Halo blight is a bacterial disease of world-wide occurrence, although some very hot, dry localities escape infection because of the lack of moisture. The pathogen is one of several bacteria which cause very similar symptoms on beans with *Xanthomonas phaseoli* causing common blight and *X. phaseoli* f. sp. *fuscans* causing fuscous blight being two other very damaging examples. Only halo blight occurs in the United Kingdom but is less important than in the United States or Africa as most of the field beans are resistant.

When leaves are invaded from the outside small, angular and necrotic spots are produced on the abaxial surface which have a wet appearance initially. Later, a halo of chlorosis develops around the area of necrosis. The bacteria produce pectolytic enzymes which digest the middle lamellae, causing tissue disintegration. On the pods, dark 'grease spot' lesions develop which are very similar in both halo blight and common blight diseases but, if bacterial exudate is present, the two can be distinguished by the yellow ooze of X. *phaseoli* or the cream ooze of P. *medicaginis*.

Seed may become infected and, if sown, give rise to systemic infections in the seedlings often resulting in death. The bacteria in the germinating seed reach the developing cotyledons in the intercellular spaces and ultimately the vascular system and both leaf lesions and stem cankers can arise from the systemic infection. Bacteria on the surfaces of cotyledons provide the inoculum for secondary cycles of infection. Stem girdling at the node above the cotyledonary attachment is common in plants originating from infected seed.

The disease is favoured by relatively cool, humid conditions. Typical symptoms will develop at 20 °C and below, whilst above 28 °C, no symptoms will develop. Common blight, by contrast, is favoured by temperature above 25 °C. The epidemic can increase rapidly by the action of rain-splash, the initial inoculum having been derived from either infected seed or from infected debris.

Control

Cultural treatments including crop hygiene, rotation and the use of clean seed provide the best form of control. Various hot water and chemical treatments can reduce seed infection but there are difficulties due to the deep-seated nature of the bacterium within the seed. Foliar sprays using copper-based compounds have given good control in New Zealand and the United States. Resistant cultivars are also available in most countries.

Bacterial Canker or Shothole of Plums and Cherries: *Pseudomonas mors-prunorum*

Bacterial canker can be a most damaging disease of stone fruits with stems, branches, leaves and fruits being attacked. The disease is described as cyclic with a well-defined winter canker stage which then alternates with the summer leaf-spotting stage.

Whilst young plum trees are particularly prone to a stem or trunk canker, cankers on cherries are usually restricted to the branches. The crotch of the tree and the angles between the branches are common sites for canker initiation but incomplete leaf scars, especially those occurring after storms in the autumn, are the most common areas with the use of too tight metal ties also affording good entry points through damage. Canker may girdle the stem causing die-back symptoms (Plate 3.2(iii)) or, less severely, the buds may fail to open and wither away. The cankers are most visible in the spring as shallow depressions of the bark. They may extend the entire length of the stem in plums, but these are less conspicuous than the cherry branch canker on which droplets of amber-coloured bacterial ooze may be seen as temperatures rise in early summer. The production of ooze is much less conspicuous on plums. However, soon after petal fall, the growth of cankers cease and the bacteria die out in the diseased tissues. From this point on, the life-cycle continues on the leaves where spotting can be seen from the end of May on cherry but a little later on plums.

The leaf spot phase begins with small, more or less circular (around 2 mm in diameter) spots of a dark brown colour. The proximity of the spots results in much coalescing leading ultimately to the falling out of the papery thin necrotic tissue in the centre of lesions giving the 'shothole' effect.

Wet weather is very conducive to the development of bacterial canker and there can be rapid increases in disease after the spread of the bacterium in wind-driven rain. Physiological races of *P. mors-prunorum* exist and there are adapted forms which attack plum or cherry with no cross-reaction.

Control

There is wide variation in susceptibility amongst cultivars of both plum and cherry although none show immunity. There are some highly resistant dwarfing rootstocks although these need to be low-worked for the highest productivity. Fungicides are recommended for cherry to try and control the leafspot phase in the spring and the autumn reinfection of the wood. The classical bordeaux mixture is still the most effective but cottonseed oil is added to the spring and midsummer sprays to reduce phytotoxic effects of the copper. Cottonseed oil is omitted from the final autumn spray, as leaf fall will follow soon after. No fungicides appear to give satisfactory control of the disease on plums.

Black Rot of Crucifers: *Xanthomonas campestris*

This is a disease which has been reported from almost all parts of the world and can be most damaging on many wild and cultivated crucifers including cabbage, cauliflower, Brussels sprouts, rape and turnip.

The early symptoms are of a localized chlorosis situated around the point of entry which may be a stomate on a cotyledon, a hydathode on the leaf edge or a site of insect damage. Within the chlorotic area, the veins become blackened and are more pronounced. The pathogen passes into the main vascular system of the stem and then may move both upwards and downwards. Early defoliation may occur because of the development of an abscission layer at the base of infected leaves. In root crops like turnip, the bacterium passes from the leaves to the fleshy root, the vascular system again showing a black discoloration which is followed by the structural breakdown of the tissue. Secondary invasion by soft-rotting bacteria (see *Erwinia carotovora* f. sp. *carotovora*) is common and the tissue is quickly reduced to a black, slimy pulp.

The bacterium can be seed-borne but it can also overwinter on infected crop debris. In either event, it is the cotyledons which are normally the first tissues attacked, the bacteria passing from the stomatal chambers, to which access is afforded from the seed coat, via the intercellular spaces to the vascular bundles. With subsequent leaf invasion, entry is through the hydathodes and then on to the vascular system. Once the infection has become established and the lesions very apparent, bacteria will be found on the surface of infected areas where they are available for dispersal by rain-splash. Seed infection occurs after the systemic invasion of the seed pods from the main plant, mildly affected seed forming the foci of infection from which the cotyledons first and then the leaves become infected.

Control

Although a two-year rotation is essential if the disease has occurred in seedbeds, the main form of control is a hot water seed treatment. In addition, it is important to use seed which is known to be disease free. In the United States, certain localities preclude the development of *X. campestris* by virtue of the low rainfall; many Pacific coastal areas are in this category and much seed is produced in these localities which can be relied upon to be healthy. There is a range of resistance within cabbages and the most resistant should be grown in areas where black rot is likely to be severe.

Diseases Caused by Viruses

Barley Yellow Dwarf Virus (BYDV)

This disease is very widely distributed and affects over 100 species of Gramineae including barley, oats, wheat and rye. From time to time, very serious outbreaks have been reported, especially on barley and oats, and it is considered to do more damage than the rusts in the United States.

The virus is aphid-borne, the most common being *Rhopalosiphum*, *Metopolophium* and *Macrosiphum* species. Symptom expression is quite conspicuous—bright yellowing of leaves on wheat or barley, or a reddening on oats which starts at the tip and progresses down the leaf blade. These symptoms appear about ten to twenty days after inoculation by the aphid and can be accompanied by some necrotic spotting, stunting, increased tillering and the production of bleached, sterile ears. If infection occurs late in the season, only the top few leaves will show symptoms.

The aphids normally require about twenty-four to forty-eight hours of feeding before they acquire the virus. Acquisition is followed by a short latent period and then the aphid is capable of transmitting the virus and can continue to do so for several weeks. In addition to the cereal crop, many wild and cultivated grasses can also act as reservoirs of infection.

Control

The onset of BYDV can be forecast by monitoring aphid populations and the most important control method is aimed at reducing aphid numbers by the application of pesticides, cypermethrin, deltamethrin, demeton-S-methyl and permethrin being used extensively. Attempts should also be made to reduce weed grasses in the headlands. Resistance has been identified in barley and many breeding programmes are now well advanced.

Maize Dwarf Mosaic Virus (MDMV)

MDMV causes serious losses in many maize-growing areas of the world. The typical primary symptom is a variegated light-green and dark-green mottling of young leaves. As plants mature, a more general chlorosis develops, often with some leaf reddening. Early infected plants are affected most both in terms of the extent of mottling and the degree of stunting. Such plants also exhibit a proliferation of adventitious buds and yield is badly affected by virtue of poor grain filling.

The virus is spread by many aphid species and is sap-transmissible. It is non-persistent and stylet-borne.

Control

It is highly recommended that resistant cultivars are grown in areas where MDMV may be a problem. The weed Johnsongrass (*Sorghum halepense*) can act as a secondary host and maize will be badly infected if planted near to this species.

Tomato Mosaic Virus (TMV)

TMV is the cause of a serious disease of solanaceous plants, including tomato and tobacco, but other species in other families can also be attacked.

The most obvious symptom is a mottle on the leaves which may vary from a slight, patchy chlorosis to bright yellow and green. The latter symptoms are often very clearly demarcated and are often referred to as 'aucuba mosaic'. Symptoms will vary considerably depending upon cultivar, strain of virus and environmental factors. There may also be some wilting in the early stages with recovery as temperatures rise; some strains of TMV produce a distortion of the leaflets, they become very narrow and fern-like, and there may be enation on the abaxial surfaces. The main stem may also exhibit a striping which will vary in colour from pale green to almost black.

Infected plants produce a wide range of fruit symptoms, the most recognizable being bronzing which make them unmarketable. The bronzing is due to a necrosis of the vascular tissue within the fruit which becomes visible through the skin, the affected areas remaining green and hard as ripening progresses. Some cultivars also have an associated dark, sunken area at the calyx end of bronzed fruit.

There are several strains of TMV and most show some degree of host cultivar specificity, although strains 0 and 1 can attack the broad range of cultivars. The disease can be carried externally and internally on the tomato seed and the virus has been identified on clothing, in cigarette tobacco and, very importantly, on debris from infected plants.

Control

Resistance against specific strains of TMV is now available in most cultivars but the gene $Tm-2^2$ gives resistance against most strains and, since its introduction, has relegated the disease to that of being relatively uncommon in most glasshouses. Heat treating seed at 70 °C for four days can eliminate virus presence except for deepseated endosperm infection. Being a contact spread disease, plants should be spaced to minimize plant contact. Crop hygiene is most important to reduce the presence of infected debris, and strict hygiene should be practised when handling the crop and, especially, with visitors from other nurseries. The crop can be protected by inoculation with a mild, harmless strain, the virus being applied under pressure using a spray gun with added carborundum to scratch the leaves and afford virus entry.

Potato Virus Diseases

Leaf Roll

One of the most important virus diseases of potato, leaf roll, is transmitted by aphids of which by far the most important is the peach potato aphid, *Myzus persicae*.

Plants often show no symptoms of this disease in the first year, but a very early infection may be expressed on the more mature plants as a rolling of the leaflets on the upper leaves. The leaves become tough and leathery, and they may show a little chlorosis or even turn a purplish colour. These are said to be the 'primary leaf roll' symptoms. It is important to realize that the virus infection is systemic and that the tubers will be infected even if very little in the way of foliage symptoms are seen.

In the second year, if infected seed is sown, the symptoms are more recognizable with a pronounced rolling of the lower leaves visible within a few weeks of emergence (Compendium Plate 4(i)). Eventually, the whole of the plant exhibits leaf-rolling and the leaves usually become thicker and feel distinctly dry and crisp and rattle when shaken. As the season progresses, the lower leaves turn brown at the margins and between the veins. The plants will look generally unthrifty with a stunted appearance which is often reflected in the size of tubers although it is almost impossible to detect infected tubers externally. Internally, there may be some vascular discoloration which is termed 'net necrosis'.

The severity of the symptoms varies considerably depending on cultivar, time of infection and amounts of virus introduced. The virus causes a degeneration of the phloem with a consequent build-up of starch in the leaves. The diversion of nutrients to the tubers cause considerable losses in tuber yield, mainly through a reduction in numbers.

Control

It is always advisable to purchase certified seed. In the United Kingdom, many localities in Scotland and Ireland are free of leaf roll virus as the growing season is too cold and windy to allow aphid multiplication and movement. Modern insecticides are also used to minimize the spread of virus via the aphid vector. Application can be as granules in the furrow, disulfoton or phorate can be used for this, or a broadcast of aldicarb will protect the early stages of crop growth and a further application of a foliar systemic such as demephion, demeton-S-methyl, or thiometon will protect until burning off. If the crop is being grown for seed, roguing is a most important operation.

Severe Mosaic

Mosaic mottling of potato leaves can be caused by one or more viruses. Severe mosaic, caused by potato virus Y, is generally the most troublesome of these diseases and is probably the most damaging of all potato virus diseases.

As with leaf roll, the symptoms vary considerably with cultivar and time of infection. With infection in the first year in some cultivars, dark brown streaks develop along the veins on the abaxial leaf surface. In other cultivars, brown spots are produced but, in either type, the leaf dies prematurely and hangs limply from the main stem by a thin, withered petiole. This symptom has been called 'leaf drop streak'.

Tubers will again be infected and the second year symptoms are very different from the previous year, the plants being very stunted with their leaves mottled yellow-green and with a slightly roughened surface. Later, the infected plants become very weak and lose their upright habit. The second year symptoms have been given the name 'rugose mosaic'.

Control

As with leaf roll, the use of certified, virus-free seed is recommended. Certain cultivars are reasonably resistant but most of the widely grown cultivars are susceptible. Aphids can acquire and transmit severe mosaic within a few minutes although, without again feeding on an infected plant, they quickly lose this ability. In consequence, insecticides are of no use to prevent entry of virus from outside the crop although, by killing aphids with the crop, they can reduce virus spread.

Mild Mosaic
The introduction of virus tested stocks has reduced this once important potato virus disease to a very minor position. No vector is known for this virus and spread is only effected by the contact of healthy leaves and infected or by machinery, animals or man. Most strains of this virus are very mild and produce little in the way of symptoms, thus posing problems in roguing operations. At most, there will be a slight mottling of the leaves but the virus produces no visible symptoms in the tubers. However, infected tubers are the main source of infection if planted. The disease only reaches importance if 'top necrosis' occurs. Small necrotic spots may be produced around the growing point and the necrosis may develop downwards, killing the plant in a few weeks.

Wheat Streak Mosaic Virus (WSMV)

This virus has a wide host range including barley, millet and oats but it is especially important on winter wheat and rye on which it causes a severe mosaic. The virus is sap-transmissible and is transmitted in the field by the adult and all nymphal stages of the wheat curl mite, *Aceria tulipae.*

The early symptoms show a yellow streaking and mottling of the leaves which result in a stunted plant, and eventually the leaves become necrotic and die. The symptoms are not as severe on maize as there is never as much necrosis but the plants may show general chlorosis. The ear husks may show a reticulate pattern of dark lines which often turns purple later. Infected maize plants are stunted with inbred lines being more susceptible than hybrids.

Control

The main control method is the use of resistant cultivars with resistance to both the virus and the vector having been detected.

Glossary

Acervulus (pl. acervuli)   A saucer-shaped, exposed fruiting structure that produces conidia from a compact layer of conidiophores, often surrounded by spines.

Aetiology   All the factors contributing to the cause of a disease.

Adult plant resistance   Resistance not expressed by seedlings, increases with plant maturity, synonymous with *mature plant resistance* (see also *field resistance*).

Aecidium   The fruiting body of rust fungi.

Aecidiospores   Spores in a chainlike series within an aecidium.

Aggressiveness   A term applied to physiological races of a pathogen that differ in the severity of their effects upon the host plant but which do not interact differentially with host cultivars. Some authorities prefer a neutral qualification of pathogenicity to describe this characteristic.

Alternate host   A second host species required by some rusts and other organisms to complete their life cycle.

Appressorium   A swelling at the tip of a fungal spore germ-tube; it is normally associated with the mechanism of adherence to the plant surface, and can be formed over the cuticle prior to direct penetration or over a stomate prior to entry via the substomatal vesicle.

Ascocarp   A mature fungal fruiting body containing asci and ascospores.

Ascospore   A spore produced within a saclike structure (the ascus).

Ascus (pl. asci)   An oval or tubular spore sac containing ascospores.

Aseptate   Without cross walls (septa).

Asexual spore   A spore produced by cell division, which is capable of developing without conjugation into a new individual.

Avirulent   Non-pathogenic.

Basidiospore   A haploid sexual spore produced on a basidium.

Basidium   A short threadlike structure, produced by germinating teliospores, which gives rise to basidiospores or sporidia; also called promycelium in the smut fungi.

Binary fission   Division of a cell into two by an apparently simple division of nucleus and cytoplasm.

Biotroph   A parasitic organism which obtains its nutrient supply only from living host tissue regardless of whether or not it can be artificially cultured.

Chlamydospore   A thick-walled, resting spore.

Chlorosis   Yellowing or whitening of the normally green tissue of plants.

Cirrhus (pl. cirrhi)   A ribbonlike column of spores extruded from the ostioles of pycnidia or perithecia.

Cleistothecium   A closed fruiting structure containing asci.

Clone   A group of identical individuals produced either by vegetative propagation or some other method or asexual reproduction or by multiplication from a single cell.

Composite mixture   A mixture of several agronomically similar lines assembled on the basis of their differences in disease resistance factors (see also *mixtures*)

Compound-interest disease   A disease which multiplies through more than one cycle within a season in a manner analogous to compound interest.

Conidiophore   A threadlike stalk upon which conidia are produced.

Conidium (pl. conidia)   Any asexual spore formed on a conidiophore.

Conjugation   The temporary union or complete fusion of two gametes or unicellular organisms.

Coremium   A sheaf-like aggregation of conidiophores.

Differential variety   A host variety, part of a set differing in disease reaction, used to identify physiologically specialized forms of a pathogen.

Dikaryon   An organism with two nuclei per cell (see *heterokaryon*).

Diploid   Having two sets of chromosomes.

Dispersing agent   A chemical added to a fungicide or bactericide formulation to aid the efficient distribution of particles of the active ingredient.

Durable resistance   Resistance that remains effective and stable during the agronomic life of a variety.

Echinulate   Having short projections on the surface of spores.

Ellipsoid   Shaped like an ellipse or oval.

Enation   A disease condition in which deformed tissue or galls develop on plant leaves or stems.

Epiphytotic   The sporadic recurrence of a disease, usually over a wide area and affecting large numbers of susceptible plants.

Exudate   Accumulation of spores or bacterial ooze.

Facultative saprophyte   A mainly parasitic organism with the ability to survive for a part of its life-cycle as a saprophyte and be cultured on artificial media.

Facultative parasite   A saprophytic organism capable of behaving as a pathogen.

Field resistance   See *adult plant resistance*

Filiform   Long and thin, threadlike.

Flagellum   Whip-like organ of motility on bacteria or zoospores.

Forma speciales   A taxon characterized from a physiological standpoint (especially host adaptation).

Fungicide   A chemical or toxin that kills or inhibits fungi.

Gall   An abnormal growth or swelling, usually caused by pathogenic organisms, nematodes, or insects.

Germ-tube   The initial hypha developing from a germinating fungal spore.

Haploid   Having one set of chromosomes.

Haustorium (pl. haustoria)   A specialized structure for extracting nutrients that is formed on some fungal hyphae following plant cell penetration.

Heterokaryon   A fungal strain or hypha in which the cells contain more than one type of nucleus per cell; the dikaryon, with two nuclei per cell, is a common example.

Heteroecious   The requirement of a pathogen for two host species to complete its lifecycle.

Heterothallic   A condition in some fungi where different thalli are required for cross-fertilization. Both male and female sex organs may be present on the same thallus but the thallus will be self-sterile.

Homoeologous   Chromosomes having in part the same sequence of genes; partly homologous.

Homokaryon   A fungal strain or hypha in which all the nuclei are of one type.

Homologous   Chromosomes with the same sequence of genes.

Homothallic   The condition in which a fungal strain is sexually self-fertile.

Honeydew   A sticky exudate (containing conidia) produced during one stage of the life cycle of *Claviceps purpurea*.

Horizontal resistance   See *race non-specific resistance*.

Hyperplasia   The enlargement of tissues by an increase in the number of cells by cell division.

Hypersensitivity   The response to attack by a pathogen of certain host plants in which the invaded cells die promptly and prevent further spread of infection.

Hypoplasia   Developmental deficiency.

Hyaline   Clear, transparent.

Hypertrophy   The enlargement of tissues by an increase in the size of the cells.

Hypha (pl. hyphae)   A tubular threadlike filament of fungal mycelium.

Immune   Not affected by pathogens.

Imperfect state   The asexual period during the life-cycle of a fungus.

Incubation period   The period between infection and the appearance of visible disease symptoms.

Infection   The entry of an organism or virus into a host and the establishment of a permanent or temporary parasitic relationship.

Infection peg   A thickening of the host cell wall in the vicinity of the penetrating hypha; lignin, callose, cellulose or suberin may be deposited at this site.

Inoculum   Spores or other disease material that may cause infection.

Isogenic lines   A series of plant lines genetically similar but which carry different specific genes for resistance to a particular pathogen.

Isolate   A culture of an organism.

Karyogamy   The fusion of nuclei in sexual reproduction.

Latent period   The period between infection and the sporulation of the pathogen on the host.

Lignituber   A thickened structure formed by the deposition of lignin surrounding the tip of the hypha penetrating a host cell and presumably functioning as a resistance mechanism.

Mature plant resistance   See *adult plant resistance*.

Meiosis   A nuclear division during sexual reproduction in which the diploid nucleus divides to produce four daughter haploid nuclei. This process, also called reduction division, is also accompanied by an exchange of genetic material, or recombination, by the process of crossing-over associated with chiasmata formation.

Mitosis   A nuclear division normally associated with cell division. The nucleus divides into two daughter nuclei with identical chromosome complements.

Mixtures   See *composite mixture*.

Monocyclic   Describes a test which measures the response of a plant or plants to a single cycle of infection by a pathogen, usually carried out by artificial inoculation (see *polycyclic*).

Monogenic   Controlled by a single gene.

Multigene variety   A variety which carries a number of specific genes governing resistance to a particular pathogen.

Multiline   A mixture of agronomically similar plant lines each of which differs genetically in terms of the factors governing resistance to disease.

Multiseptate   Having several septa (crosswalls).

Mutation   A change in the amount of structure of DNA in chromosomes which produces a discrete heritable difference compared with the 'wildtype'.

Mycelium (pl. mycelia)   A mass of hyphae that form the body of a fungus.

Mycoplasma   The smallest free-living micro-organism; it lacks a rigid cell wall and is therefore pleomorphic. Mycoplasmas cause many diseases in animals and, in plants, many diseases formerly attributed to viruses are now known to be caused by mycoplasmas.

Mycovirus   A virus that infects fungi.

Necrosis   Death of plant tissue, usually accompanied by discoloration.

Necrotroph   A fungal pathogen that causes the immediate death of the host cell as it passes through them; a colonizer of dead tissue.

Obligate parasite   A parasitic organism only capable of colonizing living host tissue.

Oligogenic resistance   Resistance controlled by one or a few genes, it should not be used as a synonym for vertical resistance.

Ostiole   A pore or opening on pycnidia or perithecia through which spores are released.

Parasitic   Describes an organism or virus which lives on another living organism, obtaining its nutrient supply from the latter but conferring no benefit in return.

Partial resistance   Resistance which is expressed by the slower development of fewer pustules or lesions compared with normally susceptible varieties.

Pathogen   An organism or virus with the capacity to cause disease in its host.

Pathogenicity   The ability of an organism or virus to cause disease.

Penetration peg   A minute protuberance from a hypha, germ-tube or appressorium which effects penetration of the host-plant surface.

Perfect stage   Sexual stage of reproduction.

Perithecium (pl. perithecia)   A closed ascocarp having an ostiole or opening.

Peritrichous   Having several flagella attached laterally, as certain bacteria.

Polar   Having flagella in the region of the end of an axis, as certain bacteria.

Polycyclic   Describes a test which measures the response of a plant or plants to repeated cycles of infection, usually carried out by natural infection under field conditions.

Promycelium   Hypha of a germinating teliospore on which basidiospores are produced.

Phylloplane   The microhabitat of the leaf surface.

Physiological race   A subdivision of a parasite species characterized by its specialization to a cultivar or different cultivars of one host species.

Phytoalexin   A substance produced in plants as a result of chemical, biological or physical stimuli and which inhibits the growth of certain micro-organisms.

Phytotoxin   Microbial metabolite producing few or no disease symptoms; it is non-specific and does not govern pathogenicity.

Pustule   A spore mass developing below the epidermis, usually breaking through at maturity.

Pycnidium   A flask-shaped or globose fungal receptacle bearing asexual spores, pycnidospores.

Pycnium   A flask-shaped fungal receptacle, characteristic of the rust fungi, bearing pycnidospores which act as spermatia.

Race   A group of organisms within a species that is distinguished by its pathogenicity.

Race non-specific resistance   Host-plant resistance which is operational against all races of a pathogen species. It is variable, sensitive to environmental changes and is usually polygenically controlled, often referred to as *horizontal resistance*.

Race-specific resistance   Host-plant resistance which is operational against one or a few races of a pathogen species; generally produces an immune or hypersensitive reaction and is controlled by one or few genes, often referred to as *vertical resistance*.

Resistance   Inherent capacity of a host plant to prevent or retard the development of disease.

Restriction enzymes   Ability to cleave double-stranded DNA at a defined nucleotide sequence.

Reticulate walls   Spore walls having a pattern of superficial lines or ridges.
Rhizoplane   The microhabitat of the root surface.
Rhizosphere   The soil immediately surrounding the root system of a plant.
Resting spore   A spore that remains dormant for a period of time before germination.
Rickettsia   Plant disease agents belonging to the Schizomycetes group of bacteria, which may cause virus-like symptoms.
Saprophyte   An organism that uses dead organic matter as food.
Sclerotium (pl. sclerotia)   A dense, compact mycelial mass capable of remaining dormant for extended periods.
Septate   Having cross-walls or septa (see also *septum*).
Septum   A cross-wall or partition.
Senescence   The phase of plant growth that extends from full maturity to death.
Serology   A technique for identifying antigens and antibodies.
Sign   Fruiting structures or other features of the pathogen that are visible on the host plant.
Simple-interest disease   A disease which goes through one cycle of multiplication only during the season, analogous to simple interest.
Somaclonal variation   Genetic variability generated within cells of a plant during cell culture.
Sorus (pl. sori)   A spore mass erupting through, or replacing, host tissue.
Sporangium (pl. sporangia)   A compact fungal mass within which spores that are usually asexual are produced.
Sporangiospore   The spores produced in a sporangium.
Spore   A minute reproductive unit in fungi and lower plant forms.
Sporidium (pl. sporidia)   The haploid sexual spore developing from a basidium; a basidio-spore.
Sporodochium (pl. sporodochia)   A cushionlike fruiting body producing conidiophores and conidia over its surface.
Sporulation   Active spore production.
Spreader   A substance added to fungicide or bactericide preparations to facilitate better contact between the spray and the sprayed surface; a surfactant.
Sticker   Added to fungicide or bactericide preparations to facilitate the adherence of the spray to the sprayed surface.
Striate   Displaying narrow parallel streaks or bands.
Stroma (pl. stromata)   A mass of mycelium from which spores develop.
Susceptible   Being subject to infection or injury by a pathogen.
Symptom   A visible response of a host plant to a pathogenic organism.
Systemic   The capacity of pathogens or chemicals to spread throughout plants, rather than remaining localized.
Transformation   A genetic engineering procedure in which the properties of cells may be modified by the insertion and expression of foreign DNA.
Transduction   The transfer of DNA or a gene or genes from a donor cell or bacterium to a recipient cell or bacterium, as by phage.
Transmission   The spread of a virus or other pathogen among individual hosts.
Tolerance   The ability to withstand infection by a pathogen without suffering undue damage or yield loss.
Tylosis   The growth of a xylem parenchyma cell through a xylem vessel wall to produce a swelling in the lumen of the vessel; tyloses are thought to impede the progress of vascular pathogens.
Urediospore   An asexual spore of the rust fungi.
Vector   An organism capable of transmitting inoculum.
Vertical resistance   See *race-specific resistance*.

Viroid A pathogenic agent, formerly classified as a virus, now known to be ribonucleic acid of low molecular weight and not a ribonucleoprotein.

Virulence The ability of an individual entity within a group of strains, *formae speciales* or isolates to cause disease under defined conditions.

Virus A nucleoprotein entity which can replicate within living cells of the host; passes through bacterium-retaining filters.

Water-soaked Appearing wet, darkened, and partially translucent.

Zoospore A fungal spore that is motile in water.

Bibliography

Text references

Agrios, G. N. (1969) *Plant Pathology*. New York, Academic Press.

Allard, R. W. (1960) *Principles of Plant Breeding*. New York and London, J. Wiley and Sons.

Allen, P. J. and Goddard, D. R. (1938) A respiratory study of powdery mildew of wheat. *American Journal of Botany*, **25**, 613–21.

Al Khesraji, T. O., Losel, D. M. and Gay, J. L. (1980) The infection of vascular tissue in leaves of *Tussilago farfara* L. by pycnial-aecial stages of *Puccinia poarum* Nich. *Physiological Plant Pathology*, **17**, 193–7.

Anonymous (1947) The measurement of potato blight. *Transactions of the British Mycological Society*, **31**, 140–1.

Anonymous (1982) *Diagnosis of Mineral Disorders in Plants*, Volume 1. HMSO, London.

Arnon, D. I. and Stout, P. R. (1939) The essentiality of certain elements in minute quantity for plants with special reference to copper. *Plant Pathology*, **14**, 371–5.

Austwick, P. K. C. (1958) Insects and the spread of fungal disease. In *Biological Aspects of the Transmission of Disease*. Horton-Smith, C. (ed), 75–9. Edinburgh, Oliver and Boyd.

Bailey, J. A. and Mansfield, J. W. (1982) *Phytoalexins*, London, Blackie.

Baker, K. F. and Cook, R. J. (1974) *Biological Control of Plant Pathogens*. Freeman, USA.

Beaumont, A. (1947) The dependence on the weather of the dates of outbreak of potato blight epidemics. *Transactions of the British Mycological Society*, **31**, 45–53.

Beever, R. E. and Byrde, R. J. W. (1982) Resistance of dicardoximide fungicides. In *Fungicide Resistance in Crop Protection*, Dekker, J. and Georgopoulos, S. G. (eds) 101–17. Wageningen, Netherlands, Centre for Agricultural Publications and Documentation.

Biffen, R. H. (1905) Mendel's laws of inheritance and wheat breeding. *Journal of Agricultural Science (Cambridge)*, **1**, 4–48.

Black, H. S. and Wheeler, H. (1966) Biochemical effects of victorin on oat tissues and mitochondria. *American Journal of Botany*, **53**, 1108–12.

Broadbent, L. (1976) Epidemiology and control of tomato mosaic virus. *Annual Review of Phytopathology*, **14**, 75–96.

Brooks, J. L., Given, J. B., Baniecki, J. F. and Young, R. J. (1974) Eradication of potato wart in West Virginia. *Plant Disease Reporter*, **58**, 291–2.

Browning, J. A. and Frey, K. J. (1981) The multiline concept in theory and practice. In *Strategies for the Control of Cereal Disease*, Jenkyn, J. F. and Plumb, R. T. (eds), 37–46. Oxford, Blackwell Scientific Publications.

Bruhn, J. A. and Fry, W. E. (1981) Analysis of potato late blight epidemiology by simulation modelling. *Phytopathology*, **71**, 612–16.

Buchanan, R. E. and Gibbons, N. E. (eds) (1974) *Bergey's Manual of Determinative Bacteriology*, 8th edition. Williams and Wilkins, Baltimore.

Bushnell, W. R. (1970) Patterns in the growth, oxygen uptake and nitrogen content of single colonies of wheat stem rust on wheat leaves. *Phytopathology*, **60**, 92–9.

Campbell, W. P. (1957) Studies on ergot infection in gramineous hosts. *Canadian Journal of Botany*, **35**, 315–20.

Carlson, R. R., Vidaver, A. K., Wysong, D. S. and Riesselman, J. H. (1979) A pressure injection device for inoculation of maize with bacterial phytopathogens. *Plant Disease Reporter*, **63**, 736.

Carver, T. L. and Carr, A. J. H. (1977) Race non-specific resistance of oats to primary infection by mildew. *Annals of Applied Biology*, **86**, 29–36.

Carver, T. L. and Griffiths, E. (1981) Relationship between powdery mildew infection, green leaf area and grain yield of barley. *Annals of Applied Biology*, **99**, 255–66.

Clarkson, D. T., Drew, M. C., Ferguson, I. B. and Sanderson, J. (1975) The effect of take-all fungus, *Gaeumannomyces graminis*, on the transport of ions by wheat plants. *Physiological Plant Pathology*, **6**, 75–84.

Clifford, B. C. (1972) The histology of race non-specific resistance to *Puccinia hordei* Orth, in barley. In *Proceedings of the European and Mediterranean Cereal Rust Conference*, **1**, 75–8.

Commonwealth Mycological Institute (1983) *Plant Pathologists Pocketbook*, 2nd edn. Kew, London, Commonwealth Agricultural Bureaux.

Cooke, B. M. and Jones, D. G. (1970) The effect of near ultra-violet irradiation and agar medium on the sporulation of *Septoria nodorum* and *Septoria tritici*. *Transactions of the British Mycological Society*, **54**, 221–26.

Cooper, R. M. (1984) The role of cell-wall degrading enzymes in infection and damage. In *Plant Diseases: Infection, Damage and Loss*, Wood, R. K. S. and Jellis, G. J. (eds), 63–72. Oxford, Blackwell Scientific Publications.

Croxall, H. E., Gwynne, D. C. and Jenkins, J. E. E. (1952) The rapid assessment of apple scab on leaves. *Plant Pathology*, **1**, 39–41.

Criukshank, I. and Perrin, D. (1963) Studies on phytoalexins. VI. The effect of some factors on its formation in *Pisum sativum* L. and the significance of pisatin in disease resistance. *Australian Journal of Biological Science*, **16**, 111–28.

Daly, J. M. (1976) Some aspects of host–pathogen interactions. In *Physiological Plant Pathology*, Heitfuss, R. and Williams, P. H. (eds). Springer-Verlag, Berlin.

Daly, J. M. (1984) The role of recognition in plant disease. *Annual Review of Phytopathology*, **22**, 273–307.

Daly, J. M. and Knoche, H. W. (1976) Hormonal movement in metabolism of host–parasite interactions. In *Biochemical Aspects of Plant–Parasite Relationships*, Friend, J. and Threlfall, D. R. (eds), 114–33. New York, Academic Press.

Daly, J. M. and Sayre, R. M. (1957) Relationship between growth and respiratory metabolism in safflower infected by *Puccinia carthami*. *Phytopathology*, **47**, 163–8.

Dazzo, F. B. (1980) Microbial adhesion to plant surfaces. In *Adsorption of Microorganisms to Surfaces*. Bitton, G. and Marshall, K. C. (eds), 311–28. New York and London, J. Wiley and Sons.

Deacon, J. W. (1975) *Phialophora radicicola* and *Gaeumannomyces graminis* on roots of grasses and cereals. *Transactions of the British Mycological Society*, **61**, 471–85.

Défago, G. (1977) Role des saponines dans la résistance des plantes aux maladies fongiques. *Berichte Schweizerischen Botanischen Gesellschaft*, **87**, 79–132.

Dekker, J. (1981) Impact of fungicide resistance on disease control. *Proceedings of the 1981 British Crop Protection Conference*, 857–64.

Doling, D. A. (1965) Single-bath hot-water treatment for the control of loose smut (*Ustilago nuda* (Jens.) Rostr.) in cereals. *Annals of Applied Biology*, **55**, 295–301.

Dowson, W. J. (1957) *Plant Diseases Due to Bacteria*. Cambridge, Cambridge University Press.

Ellis, M. A., Madden, L. V. and Wilson, L. L. (1984) Evaluation of an electronic apple scab predictor for scheduling fungicides with curative activity. *Plant Disease*, **68**, 1055–7.

Evans, E., Couzens, B. J. and Griffiths, W. (1965) Timing experiments on the control of potato blight with copper fungicides in the United Kingdom. *World Review of Pest Control*, **4**, 84–92.

Ferhmann, H. and Schrodter, H. (1974) Okologische Untersuchungen Zur Epidemiologie von *Cercosporella herpotrichoides*. *Phytopathologische Zeitschrift*, **74**, 161–75.

Fitt, B. D. L. and Hornby, D. (1978) Effects of root-infecting fungi on wheat transport processes and growth. *Physiological Plant Pathology*, **13**, 335–46.

Flor, H. H. (1956) The complementary genic systems in flax and flax rust. *Advances in Genetics*, **8**, 29–54.

Fox, R. A. (1977) The impact of ecological, cultural and biological factors on the strategy and costs of controlling root diseases in tropical crops as exemplified by *Hevea brasilensis*. *Journal of the Rubber Research Industry, Sri Lanka*, **54**, 329–62.

Fox, R. T. V., Manners, J. G. and Myers, A. (1971) Ultrastructure of entry and spread of *Erwinia carotovora* var. *atroseptica* into potato tubers. *Potato Research*, **141**, 61–73.

Fry, W. E. (1982) *Principles of Disease Management*. New York, Academic Press.

Gäumann, E. (1950) *Principles of plant infection*. London, Crosby Lockwood.

Gäumann, E. (1957) Fusaric acid as a wilt toxin. *Phytopathology*, **47**, 342–57.

Gibbs, A. and Harrison, B. D. (1976) *Plant Virology: The Principles*. London, Edward Arnold.

Gregory, P. H. (1973) *The Microbiology of the Atmosphere*. 2nd edn. London, Leonard Hill.

Griffiths, E. and Hann, C. Ao. (1976) Dispersal of *Septoria nodorum* spores and spread of glume blotch of wheat in the field. *Transactions of the British Mycological Society*, **67**, 413–18.

Griffiths, H. M., Jones, D. G. and Akers, A. (1985) A bioassay for predicting the resistance of wheat leaves to *Septoria nodorum*. *Annals of Applied Biology*, **107**, 293–300.

Groenewegen, L. J. M. and Zadoks, J. C. (1979) Exploiting within-field diversity as a defence against cereal diseases: a plea for 'poly-genotype' varieties. *Indian Journal of Genetics and Plant Breeding*, **39**, 81–94.

Habeshaw, D. (1984) Effects of pathogens on photosynthesis. In *Plant Diseases: Infection, Damage and Loss*. Wood, R. K. S. and Jellis, G. J. (eds), 63–72. Oxford, Blackwell Scientific Publications.

Hammerschmidt, R. and Kuc, J. (1979) Isolation and identification of phytuberin from *Nicotiana tabacum* previously infiltrated with an incompatible bacterium. *Phytochemistry*, **18**, 874–5.

Hart, H. (1929) Relation of stomatal behaviour to stem-rust resistance in wheat. *Journal of Agricultural Research*, **39**, 929–48.

Heitfuss, R. (1965) Untersuchungen zur Physiologie des Temperaturgesteurten Vertraglichkeitsgrades von Weizen und *Puccinia graminisritici*. *Phytopathologische Zeitschrift*, **54**, 379–400.

Hirst, J. M. and Hurst, G. W. (1967) Long distance spore transport: vertical sections of spore clouds over the sea. *Journal of General Microbiology*, **48**, 357–77.

Hislop, E. C. and Stahmann, M. A. (1971) Peroxidase and ethylene production by barley leaves infected by *Erysiphe graminis* f. sp. *hordei*. *Physiological Plant Pathology*, **1**, 297–312.

Hollins, M. (1965) Disease control through virus-free stock. *Annual Review of Phytopathology*, **3**, 367–96.

Hornby, D. and Fitt, B. D. L. (1982) Effects of root-infecting fungi on structure and function of cereals. In *Effects of Disease on the Physiology of the Growing Plant*. Ayres, P. G. (ed), 101–30. Cambridge, Cambridge University Press.

Ingold, C. T. (1971) *Fungal Spores, Their Liberation and Dispersal*. Oxford, Oxford University Press.

International Atomic Energy Authority (1977) *The Use of Induced Mutations for Improving Disease Resistance in Plants*. IAEA, Vienna.

James, W. C. (1971) A manual of assessment keys for plant diseases. *Canadian Department of Agriculture Publication*, 1458.

James, W. C. (1974) Assessment of plant diseases and losses. *Annual Review of Phytopathology*, **12**, 27–48.

Jeger, M. J., Griffiths, E. and Jones, D. G. (1983) Seasonal variation in the components of partial resistance of seedlings of winter wheat cultivars Maris Huntsman and Maris Ranger to *Septoria nodorum*. *Plant Pathology*, **32**, 187–96.

Jellis, G. J., Gunn, R. E. and Boulton, R. E. (1984) Variation in disease resistance among potato somaclones. In *Abstracts of Conference Papers, Triennial, Conference, European Association of Potato Research, Interlaken*. Winiger, F. A. (ed), 380–1. Wageningen, European Association of Potato Research.

Johnson, R., and Taylor, A. J. (1972) Isolates of *Puccinia striiformis* collected in England from the wheat varieties Maris Beacon and Joss Cambier. *Nature (London)*, **238**, 105–6.

Johnson, R. and Law, C. N. (1973) Cytogenetic studies on the resistance of the wheat variety Bersee to *Puccinia striiformis*. *Cereal Rusts Bulletin*, **1**, 38–43.

Johnson, R. and Allen, D. J. (1975) Induced resistance to rust diseases and its probable role in the resistance of multiline varieties. *Annals of Applied Biology*, **80**, 359–64.

Jones, D. G. and Clifford, B. C. (1983) *Cereal Diseases: Their Pathology and Control*. London, John Wiley.

Kable, P. F. and Jeffrey, H. (1980) Selection for tolerance in organisms exposed to sprays of biocide mixtures: a theoretical model. *Phytopathology*, **70**, 8–12.

Kerr, A. and Htay, K. (1974) Biological control of crown gall through bacteriocin production. *Physiological Plant Pathology*, **4**, 37–44.

Knott, D. R. and Dvorak, J. (1976) Alien germ plasm as a source of resistance to disease. *Phytopathology*, **64**, 211–35.

Kojima, M. and Uritani, I. (1976) Possible involvement of furanoterpenoid phytoalexins in establishing host–parasite specificity between potato and various strains of *Ceratocystis fimbriata*. *Physiological Plant Pathology*, **8**, 97–111.

Kolattukudy, P. E. (1985) Enzymatic penetration of the plant cuticle by fungal pathogens. *Annual Reviews of Phytopathology*, **23**, 223–50.

Krause, R. A., Massie, L. B. and Hyre, R. A. (1975) Blitecast, a computerized forecast of potato blight. *Plant Disease Reporter*, **59**, 95–8.

Kuc, J. (1972) Phytoalexins. *Annual Reviews of Phytopathology,* 10, 207–32.

Kurasawa, E. (1926) *Journal of the Natural History Society of Formosa,* 16, 213. Abstract in *Biological Abstracts,* 48, 1066 (1929).

Lancashire, P. D. and Jones, D. G. (1985) Components of partial resistance to *Septoria nodorum* in winter wheat. *Annals of Applied Biology,* 106, 541–53.

Large, E. C. (1940) *The Advance of the Fungi.* Jonathan Cape, London.

Large, E. C. (1952) The interpretation of progress curves for potato blight and other plant diseases. *Plant Pathology,* 1, 109–17.

Large, E. C. (1954) Growth stages in cereals: illustrations of the Feekes scale. *Plant Pathology,* 3, 128–9.

Large, E. C. (1958) Losses caused by potato blight in England and Wales. *Plant Pathology,* 7, 39–48.

Large, E. C. and Doling, D. A. (1962) The measurement of cereal mildew and its effect upon yield. *Plant Pathology,* 11, 47–57.

Larkin, P. J. and Scowcroft, W. R. (1981) Somaclonal variation—a novel source of variability from cell cultures for plant improvement. *Theoretical and Applied Genetics,* 60, 197–214.

Leach, C. M. (1962) The sporulation of diverse species of fungi under near ultra-violet radiation. *Canadian Journal of Botany,* 40, 151–61.

Leath, K. T. and Stewart, D. M. (1966) A method for the separation of isolates of *Puccinia graminis. Plant Disease Reporter,* 50, 312.

Link, K. P., Dickson, A. D. and Walker, J. C. (1929) Further observations on the occurrence of protocatechuic acid in pigmented onion scales and its relation to disease resistance in the onion. *Journal of Biological Chemistry,* 84, 719–25.

Lupton, F. G. H. and Macer, R. C. F. (1962) Inheritance of yellow rust (*Puccinia glumarum* Erikss. and Henn.) in some varieties of wheat. *Transactions of the British Mycological Society,* 45, 21–45.

Maddison, A. C. and Manners, J. G. (1972) Sunlight and viability of cereal rust uredospores. *Transactions of the British Mycological Society,* 59, 429–43.

Maddison, A. C. and Manners, J. G. (1973) Lethal effects of artificial ultra-violet radiation on cereal rust uredospores. *Transactions of the British Mycological Society,* 60, 471–94.

Mansfield, J. W. and Deverall, B. J. (1974) Changes in wyerone acid concentrations in leaves of *Vicia faba* after infection by *Botrytis cinerea* or *B. fabae. Annals of Applied Biology,* 77, 227–35.

Mantle, P. G. and Shaw, S. (1977) A case study of the aetiology of ergot disease of cereals and grasses. *Plant Pathology,* 26, 121–26.

Marsh, R. W. (ed.) (1977) *Systemic Fungicides,* 2nd edn. London, Longman.

Martin, H. and Worthing, C. R. (1976) *Insecticide and Fungicide Handbook,* 5th edn. Oxford, Blackwell.

Mathews, G. A. (1979) *Pesticide Application Methods.* London: Butler and Tanner.

Mathews, R. E. F. (1979) Classification and nomenclature of viruses. 3rd report of the International Committee on Taxonomy of Viruses. *Intervirology,* 12, 131–296.

Mathews, R. E. F. (1981) *Plant Virology.* London, Academic Press.

Matthyse, A. G. and Curlitz, R. H. G. (1982) Plant cell range for attachment of *Agrobacterium tumefaciens* to tissue culture cells. *Physiological Plant Pathology,* 21, 381–7.

Mayama, S., Matsurura, Y., Iida, H. and Tani, T. (1982) The role of avenalumin in the resistance of oat to crown rust, *Puccinia coronata* f. sp. *avenae. Physiological Plant Pathology,* 20, 189–99.

Melander, L. W. and Craigie, J. H. (1927) Nature of resistance of *Berberis* spp. to *Puccinia graminis. Phytopathology,* 17, 95–114.

Mills, W. D. and Laplante, A. A. (1954) Diseases and insects in the orchard. *New York Agricultural Experimental Station (Ithaca) Extension Bulletin,* 711, 21–7.

Moseman, J. G. (1966) Genetics of powdery mildews. *Annual Review of Phytopathology,* 4, 269–90.

Muller, K. O. and Borger, H. (1940) Experimentalle Untersuchungen uber die *Phytophthora*—Resistenz de Kartoffel—zugleich ein Beitrag zum Problem de erworbenen Resistenz in Pflanzenreich. *Arbeiten aus der Biologischen Abteilung (Anstalt-Reichsanst,* Berlin, 23, 189–231.

Murai, N., Sutton, D. W., Murray, M. G., Slighton, J. L., Merlo, D. J. *et al.* (1983) Phaseollin gene from bean is expressed after transfer of sunflower via tumour-induced plasmid vectors. *Science,* 222, 476–82.

Newton, R., Lehmann, J. V. and Clarke, A. E. (1929) *Canadian Journal of Research,* 1, 5; Abstract in *Review of Applied Mycology,* 9, 95, (1930).

Noble, M., MacGarvie, Q. D., Hams, A. F. and Leafe, E. L. (1966) Resistance to mercury of *Pyrenophora avenae* in Scottish seed oats. *Plant Pathology,* 15, 23–8.

Nutman, F. J. and Roberts, F. M. (1960) Investigations on a disease of *Coffea arabica* caused by a form of *Colletotrichum coffeanum* Noack. II. Some factors affecting germination and infection, and their relation to disease distribution. *Transactions of the British Mycological Society,* 43, 643–59.

Ogawa, J. W., Gilpatrick, J. D. and Chiarappa, L. (1977). Review of plant pathogens resistant to fungicides and bactericides. *FAO Plant Protection Bulletin,* 25, 97–111.

Owera, S. A. P., Farrar, J. F. and Whitbread, R. (1981) Growth and phytosynthesis in barley infected with brown rust. *Physiological Plant Pathology*, **18**, 79–90.

Parlevliet, J. (1979) Components of resistance that reduce the rate of epidemic development. *Annual Review of Phytopathology*, **69**, 281–6.

Pegg, G. F. (1984) The role of growth regulators in plant disease. In *Plant Disease: Infection, Damage and Loss.* Wood, R. K. S. and Jellis, G. K. (eds), 29–48. Oxford, Blackwell Scientific Publications.

Pegg, G. F. (1985) Life in a black hole—the micro-environment of the vascular pathogen. *Transactions of the British Mycological Society*, **85**, 1–20.

Polley, R. W. and King, J. E. (1973) A preliminary proposal for the detection of barley mildew infection periods. *Plant Pathology*, **22**, 11–16.

Priestley, R. H. (1978) Detection of increased virulence in populations of yellow rust. In *Plant Disease Epidemiology:* Scott, P. R. and Bainbridge, A. (eds). Oxford, Blackwell Scientific Publications.

Pueppke, S. G. and Van Etten, H. D. (1976) Accumulation of pisatin and three additional antifungal pterocarpans in *Fusarium solani*—infected tissues of *Pisum sativum*. *Physiological Plant Pathology*, **8**, 51–61.

Puranik, S. B. and Mathre, D. E. (1971) Biology and control of ergot on male sterile wheat and barley. *Phytopathology*, **61**, 1075–80.

Purdy, R. E. and Kolattukudy, P. E. (1976) Hydrolysis of plant cuticle by plant pathogens. Purification, amino-acid composition and molecular weight of two isozymes of cutinase and a non-specific esterase from *Fusarium solani* f. *pisi*. *Biochemistry*, **14**, 2824–31.

Raymundo, S. A. and Young, H. C. (1974) Improved methods for the axenic culture of *Puccinia recondita* f. sp. *tritici*. *Phytopathology*, **64**, 262–3.

Ride, J. P. and Drysdale, R. B. (1972) A rapid method for the chemical estimation of filamentous fungi in plant tissue. *Physiological Plant Pathology*, **2**, 7–15.

Riley, R., Chapman, V. and Johnson, R. (1968) The incorporation of alien disease resistance in wheat by genetic interference with the regulation of meiotic chromosome synapsis. *Genetic Research*, **12**, 199–219.

Rishbeth, J. (1950) Observations on the biology of *Fomes annosus* with particular reference to East Anglian pine plantations. III. Natural and experimental infection of pines and some factors affecting severity of the disease. *Annals of Botany*, **15**, 221–46.

Roberts, F. M. (1946) Underground spread of potato virus X. *Nature (London)*, **158**, 663.

Romig, R. W. and Caldwell, R. M. (1964) Stomatal exclusion of *Puccinia recondita* by wheat peduncles and sheaths. *Phytopathology*, **54**, 214–18.

Russell, G. E. (1978) *Plant Breeding for Pest and Disease Resistance*. London, Butterworths.

Salaman, R. N. (1911) Studies in potato breeding. In *Proceedings of the 4th International Congress of Genetics*, p. 397.

Schafer, J. F. (1971) Tolerance to plant disease. *Annual Review of Phytopathology*, **9**, 235–52.

Scott, K. J. (1965) Respiratory enzymic activities in the host and pathogen of barley leaves infected with *Erysiphe graminis*. *Phytopathology*, **55**, 438–41.

Scott, K. J. and Smillie, R. M. (1966) Metabolic regulation in diseased leaves. I. The respiratory rise in barley leaves infected with powdery mildew. *Plant Physiology*, **41**, 289–97.

Scott, P. R. and Hollins, T. W. (1974) Effects of eyespot on the yield of winter wheat. *Annals of Applied Biology*, **78**, 269–79.

Seevers, P. M. and Daly, J. M. (1970) Studies on wheat stem rust resistance controlled at the Sr6 locus. II. Peroxidase activities. *Phytopathology*, **54**, 1642–7.

Sequeira, L. (1963) Synthesis of indoleacetic acid by *Pseudomonas solanacearum*. *Phytopathology*, **54**, 1240–6.

Sequeira, L. (1978) Lectins and their role in host-pathogen specificity. *Annual Reviews of Phytopathology*, **16**, 453–81.

Shayk, M., Soliday, C. L. and Kolattukudy, P. E. (1977) Proof for the production of cutinase by *Fussarium solani* f. *pisi* during penetration into its host, *Pisum sativum*. *Plant Pathology*, **60**, 170–2.

Shephard, J. F., Bidney, D. and Shahin, E. (1980) Potato protoplasts in crop improvement. *Science*, **208**, 17–24.

Smedegård-Petersen, V. (1980) Increased demand for respiratory energy of barley leaves reacting hypersensitively against *Erysiphe graminis*, *Pyrenophora teres* and *Pyrenophora graminea*. *Phytopathologie Zeitschrift*, **99**, 54–62.

Smedegård-Petersen, V. (1984) The role of respiration and energy generation in diseased and disease-resistant plants. In *Plant Disease: Infection Damage and Loss.* Wood, R. K. S. and Jellis, G. J. (eds), 73–85. Oxford, Blackwell Scientific Publications.

Smidt, M. L. and Kosuge, T. (1978) The role of indole-3-acetic acid accumulation by alpha methyl tryptophan resistant mutants of *Pseudomonas savastanoi* in gall formation on oleanders. *Physiological Plant Pathology*, **13**, 203–14.

Smith, L. P. (1961) The duration of surface wetness (a new approach to horticultural climatology). *Report of the XVth International Horticultural Congress* (Nice).

Stakman, E. C. and Harrar, J. G. (1957) *Principles of Plant Pathology*. New York, Ronald Press.

Staub, T. and Sozzi, D. (1984) Fungicide resistance. *Plant Disease*, **68**, 1026–31.

Steadman, J. R. and Sequeira, L. (1970) Abscissic acid in tobacco plants: tentative identification and its relation to stunting induced by *Pseudomonas solanacearum*. *Plant Pathology*, **45**, 691–7.

Tarr, S. A. J. (1972) *The Principles of Plant Pathology*. London, Macmillan.

Teng, P. S., Blackie, M. J. and Close, R. C. (1980) Simulation of the barley leaf rust epidemic: structure and validation of BARSIM—I. *Agricultural Systems*, **5**, 55–73.

Teng, P. S., Close, R. C. and Blackie, M. J. (1979) Comparison of models for estimating yield loss caused by leaf rust (*Puccinia hordei*) on Zephyr barley in New Zealand. *Phytopathology*, **69**, 1239–44.

Tervet, I. W., Rawson, A. J., Cherry, L. and Saxon, R. B. (1951) A method for the collection of microscopic particulars. *Phytopathology*, **41**, 282–5.

Tomlinson, J. A., Carter, A. L., Dale, W. T. and Simpson, C. J. (1970) Weed plants as sources of cucumber mosaic virus. *Annals of Applied Biology*, **66**, 11–16.

Tottman, D. R. and Makepeace, R. J. (1979) An explanation of the decimal code for the growth stages of cereals, with illustrations. *Annals of Applied Biology*, **93**, 221–234.

Turner, E. M. C. (1961) An enzymic basis for pathogenic specificity in *Ophiobolus graminis*. *Journal of Experimental Botany*, **12**, 169–75.

Turner, J. G. (1984) Role of toxins in plant disease. In *Plant Disease: Infection, Damage and Loss*. Wood, R. K. S. and Jellis, G. J. (eds) 1–12. Oxford, Blackwell Scientific Publications.

Van der Plank, J. E. (1963) *Plant Diseases: Epidemics and Control*. New York, Academic Press.

Van der Plank, J. E. (1967) Epidemiology of fungicidal action. In *Fungicides: An Advanced Treatise*. Torgeson, D. C. (ed) **1**, 63–92. New York, Academic Press.

Van der Wal, A. F. and Zadoks, J. C. (1976) Towards mass production of uredospores of brown rust on wheat. *Cereal Rusts Bulletin*, **4**, 9–12.

Van Etten, H. D. and Pueppke, S. G. (1976) Isoflavonoid phytoalexins. *Annal Review of Phytochemical Society*, **13**, 239–89.

Varns, J. and Kuc, J. (1971) Suppression of rishitin and phytuberin accumulation and hypersensitive response in potato by compatible races of *Phytophthora infestans*. *Phytopathology*, **61**, 178–81.

Vasil, I. K. (1980) Androgenetic haploids. *International Review of Cytology*, Supplement 11A, 195–223.

Waggoner, P. E., Horsfall, J. G. and Lukens, R. J. (1972) EPIMAY, a simulator of southern corn leaf blight. *Bulletins of the Connecticut Agricultural Experimental Station*, **279**, 84 pp.

Walker, J. C. (1969) *Plant Pathology*, 3rd edn. New York, McGraw-Hill.

Wakimoto, S. and Yoshi, H. (1958) *Annals of the Phytopathological Society of Japan*, **23**, 79. Abstract in *Review of Applied Mycology*, **38**, 142 (1959).

Warcup, J. H. (1950) The soil-plate method for isolation of fungi from soil. *Nature* (*London*), **166**, 117–8.

Watson, M. A. (1966) The relation of annual incidence of beet yellowing viruses in sugar beet to variation in weather. *Plant Pathology*, **15**, 145–9.

Wenzel, G. (1985) Strategies for unconventional breeding for disease resistance. *Annual Reviews of Phytopathology*, **23**, 149–72.

Whitney, P. J. (1976) *Microbial Plant Pathology*. London, Hutchinson.

Williams, W. (1960) *Genetical Principles and Plant Breeding*. London, Davis.

Wolfe, M. S. (1985) The current status and prospects of multiline cultivars and variety mixtures for disease resistance. *Annual Reviews of Phytopathology*, **23**, 251–73.

Wood, R. K. S. and Graniti, A. (eds) (1976) *Specificity in Plant Diseases*. New York, Plenum.

Wood, R. K. S. and McCrae, S. I. (1979) Synergism between enzymes involved in the solubilization of native cellulose. In *Advances in Chemistry*, Volume 181. Hydrolysis of cellulose: Mechanisms of enzymatic and acid catalysts. Brown, R. D. and Juraski, L. (eds), 181–209. American Chemical Society.

Zadoks, J. C. (1981) EPIPRE: a disease simulator and pest management system for winter wheat developed in the Netherlands. *EPPO Bulletin*, **11**, 365–9.

Zadoks, J. C., Chang, T. T. and Konzak, C. F. (1974) A decimal code for the growth stages of cereals. *Weed Research*, **14**, 415–21.

References for general background reading

Agrios, G. N. (1969) *Plant Pathology*. New York, Academic Press.

Bailey, J. A. and Mansfield, J. W. (1982) *Phytoalexins*, London, Blackie.

Commonwealth Mycological Institute (1983) *Plant Pathologists Pocketbook* 2nd edn. Kew, London, Commonwealth Agricultural Bureaux.

Day, P. R. (1974) *Genetics of Host-Parasite Interaction.* San Francisco, Freeman.

Deacon, J. W. (1984) *Introduction to Modern Mycology,* 2nd edn. Oxford, Blackwell Scientific Publications.

Diagnosis of Mineral Disorders in Plants, Volume 1 (1982) London, HMSO.

Dickinson, C. H. and Lucas, J. A. (1977) *Plant Pathology and Plant Pathogens.* Basic Microbiology Volume **6.** Oxford, Blackwell Scientific Publications.

Fry, W. E. (1982) *Principles of Plant Disease Management.* New York, Academic Press.

Garrett, S. D. (1970) *Pathogenic Root-infecting Fungi.* Cambridge, Cambridge University Press.

Gibbs, A. and Harrison, B. D. (1976) *Plant Virology: the Principles.* London, Edward Arnold.

Ingold, C. T. (1971) *Fungal Spores, their Liberation and Dispersal.* Oxford, Oxford University Press.

Jenkyn, J. F. and Plumb, R. T. (1981) (eds) *Strategies for the Control of Cereal Disease.* Oxford, Blackwell Scientific Publications.

Jones, D. Gareth and Clifford, B. C. (1983) *Cereal Diseases: Their Pathology and Control.* London, John Wiley.

Manners, J. G. (1982) *Principles of Plant Pathology.* Cambridge, Cambridge University Press.

Mathews, G. A. (1979) *Pesticide Application Methods.* London, Butler and Tanner.

Russell, G. E. (1978) *Plant Breeding for Pest and Disease Resistance.* London, Butterworths.

Scott, P. R. and Bainbridge, A. (1978) (eds) *Plant Disease Epidemiology.* Oxford, Blackwell Scientific Publications.

Talbot, P. H. B. (1971) *Principles of Fungal Taxonomy.* London, Macmillan.

Van der Plank, J. E. (1963) *Plant Diseases: Epidemics and Control.* New York, Academic Press.

Walker, J. C. (1969) *Plant Pathology,* 3rd edn. New York, McGraw-Hill.

Walkey, D. G. A. (1985) *Applied Plant Virology.* London, William Heinemann.

Webster, J. (1983) *Introduction to Fungi,* 2nd edn. Cambridge, Cambridge University Press.

Wheeler, B. E. J. (1969) *Introduction to Plant Disease.* London, John Wiley.

Wheeler, H. (1975) *Plant Pathogenesis.* Berlin, Springer-Verlag.

Whitney, P. J. (1976) *Microbial Plant Pathology.* London, Hutchinson.

Wood, R. K. S. and Jellis, G. J. (1984) (eds) *Plant Diseases—Infection, Damage and Loss.* Oxford, Blackwell Scientific Publications.

Readers are also recommended to read review articles which appear in the *Annual Review of Phytopathology.*

Index